THE END OF TIME

Adrian Gilbert is one of the top writers in the field of ancient mysteries. His book with Robert Bauval, *The Orion Mystery: Unlocking the Secrets of the Pyramids*, was a huge international success, while *The Mayan Prophecies: Unlocking the Secrets of a Lost Civilization* sold over a million copies worldwide.

www.adriangilbert.co.uk

THE END OF TIME

THE MAYAN PROPHECIES REVISITED

ADRIAN GILBERT

MAINSTREAM
PUBLISHING

EDINBURGH AND LONDON

**To my daughter Alice, of the generation who
will see what happens to the world in
decades long after the end of time**

This edition published, 2007

First published in Great Britain in 2006 by
MAINSTREAM PUBLISHING COMPANY
(EDINBURGH) LTD
7 Albany Street
Edinburgh EH1 3UG

ISBN 9781845962609

A catalogue record for this book is available
from the British Library

Typeset in Stone Informal and Frutiger

Printed in Great Britain by
Cox & Wyman Ltd, Reading

Picture Credits

All plates are the copyright of Adrian and Dee Gilbert.

All figures are the copyright of Adrian Gilbert, except those listed below, which have been taken from the following sources:

Page 37: portrait of Bernadino Sahagun, from Sahagun, *A History of Ancient Mexico*, translated by Fanny R. Bandelier from the Spanish version of Carlos Maria Bustamente.

Page 74: first page of the *Fejérváry-Mayer* codex.

Page 91: Figure 8, from Stephens and Catherwood, *Incidents of Travel in Central America, Chiapas and Yucatán*, Vol I.

Page 92: Figure 9, from Stephens and Catherwood, *Incidents of Travel in Central America, Chiapas and Yucatán*, Vol II.

Pages 100, 101: Figures 10, 11, from Morley, *An Introduction to the Study of the Maya Hieroglyphs*.

Page 116: Figure 14, from Stephens and Catherwood, *Incidents of Travel in Central America, Chiapas and Yucatán*, Vol I.

Page 117: Figure 15, from Stephens and Catherwood, *Incidents of Travel in Central America, Chiapas and Yucatán*, Vol I.

Page 131: Figure 18, from Wägner and Macdowall, *Asgard and the Gods*.

Page 134: Figure 19, after Taube, *Aztec and Maya Myths*.

Page 138: Figure 21, after Wauchope, *Modern Maya Houses*.

Page 144: Figure 24, from Stephens and Catherwood, *Incidents of Travel in Central America, Chiapas and Yucatán*, Vol II.

Page 145: Figure 25, from Stephens and Catherwood, *Incidents of Travel in Central America, Chiapas and Yucatán*, Vol II.

Page 146: Figure 26, from Maudsley, *Archaeology, Biologia Centrali Americana*.

Page 163: Figure 31, from the *Codex Mendoza*.

Page 173: Figure 36, from the *Dresden Codex*, drawing by Carlos Villacorta.

Page 197: Figure 38, from Fell, *America BC*, after Peet, *The Mound Builders*, 1892.

Page 239: Figure 43, from Bolio, *The Geometry of the Maya*.

Page 249: Figure 44, from Stephens and Catherwood, *Incidents of Travel in Yucatán*, Vol II.

Page 272: Figure 46, after Taube, *Aztec and Maya Myths*.

Page 288: Figure 47, from Stephens and Catherwood, *Incidents of Travel in Central America, Chiapas and Yucatán*, Vol II.

Acknowledgements

The first thank you is to my wife, Dee, for putting up with having our house invaded by Aztecs and Maya for the second time in ten years. Her patience is much appreciated. Thanks to Dr Charles Thomas Cayce, to John van Auken and Ken Skidmore. Without them, this book would never have happened. Thanks too to Greg and Lora Little, who have been kind enough to share their discoveries concerning remnants of lost Atlantis in the Bahamas. I look forward to more conversations with them on this and other subjects. Thanks also to the late José Diaz Bolio, rediscoverer of the rattlesnake cult. His perseverance in the face of obstacles has been an inspiration. I hope that, in recompense, I have in some small way brought his work to the attention of the public. A large thank you to Carlos Barrios and the Mayan elders, who have done and are doing so much to keep alive the light of ancient wisdom in the twilight years of this current age. To say that witnessing Carlos conduct a fire ceremony within the sacred precincts of an ancient Mayan city was inspiring is an understatement. For me, it made the whole subject of Mayan prophecies come alive. Thanks, too, to Mitch Battros, who, with his ECTV radio show, has brought the concept of sunspot cycles and their effect on global warming to the attention of experts as well as amateurs.

A big thank you to my British publishers, Mainstream, for continuing to support the 'ancient mysteries' genre when so many others have walked away from it; also to my US publishers, the Association for Research and Enlightenment, for being so patient in waiting for this book to materialise.

Thanks also to the University of Oklahoma Press, Bear & Co., Area-Maya, Fisk University Press, Kingsport Press, Dover Books, Eagle Wing Books, Simon & Schuster, the Academy of Future Science and all other relevant publishers for the use of quotations from their books.

Finally, thanks to all the individuals I have met 'on the way' who have, often unknowingly, influenced my thinking. Without your input, I could not do my work.

Contents

Preface

According to the ancient Maya, our present 'Age of the Jaguar', which began on 13 August 3114 BC, will reach its completion on 21 December 2012. This 'end-date' has intrigued scholars ever since the Mayan system of timekeeping was rediscovered about 100 years ago. It appears all the more special once it is realised that it is not an arbitrary date but reflects certain astronomical facts. On the next day, 22 December 2012, the Sun will be aligned, at the winter solstice, with a 'star-gate' at the centre of our galaxy.[1] As it only does this every 25,800 years, this is the first time in recorded history that such an event will be witnessed. This poses the obvious question: why did the ancient Maya, a Stone Age people who didn't use wheels, let alone telescopes, invent a calendar with its predicted climax a unique astronomical event that, as far as they were concerned, would happen thousands of years in the future?

This is not the only mystery. Equally strange is that, according to many inscriptions that they left behind, the start date of their calendar, a named day equivalent to 13 August 3114 BC on our Gregorian calendar, was some 3,000 years before their own times. It is even 2,000 years before the 'Olmec', the earliest recognised civilisation in Central America, thought to have been a precursor of the Maya.

This being so, one has to ask what, if anything, was so special about 13 August 3114 BC? Why did the Maya claim that it was on this very day that the 'gods' inaugurated the present age, and did they know that 13 *baktuns*, or 1,872,000 days, later the Sun would be positioned at a stargate on the winter solstice?

These are questions for which academic archaeology currently has no answers. However, very curiously, the epoch of c. 3100, which corresponds with the start of the present phase of the Mayan calendar, is seen worldwide as having been important. In Britain, it marks the early phase of erecting astronomically aligned monuments, most notably Stonehenge I. In Africa, it corresponds to the beginning of Dynastic Egypt; in Mesopotamia, to the development of the cuneiform script; in America, to the first cultivation of maize. Taken separately, these events are each, in their own way, significant steps on the path of world civilisation. Together, they form the basis of a cultural revolution that in different ways seems to have influenced the whole planet.

However, this global view leaves unanswered a further question: what motivated diverse peoples, at roughly the same time on different parts of planet Earth, to suddenly begin doing things differently? Or to put it another way: where did the idea of civilisation come from? Contemporary science tells us that civilisation is self-creating, the result of cultural evolution. However, while this may be part of the truth, I do not believe it to be the whole truth. For when we examine the writings, art and religion of ancient peoples, it invariably tells us the same story: there were civilisations earlier than our own which were brought to an end by barely remembered catastrophes.

In this book, I propose to bring the reader up to speed on the subject of Central American wisdom and to show how the behaviour of this star- and calendar-obsessed culture was governed by one principle: that events repeat themselves. For unlike us, who have lived through the great technological revolutions of the nineteenth and twentieth centuries, they did not think in terms of progress. They saw time as cyclical and therefore believed that it was the natural order of things that each age should go through a cycle that ended in the destruction of all that had been achieved. This, however, would not be a permanent end but rather the beginning of a new cycle. For this reason, they studied time-cycles, long and short, as a method of fortune-telling. Essentially fatalistic in their beliefs, they needed to know where they stood in relation to time as a means of coming to terms with life.

Unfortunately, knowing this still doesn't answer the question of where this obsession with time and cycles originated. However, as a strange parallelism exists between certain Mayan cosmological myths and similar myths from elsewhere, the possibility of pre-Columbian contact between America and the rest of the world – and therefore with the ancient wisdom of Europe and Africa – has to be a possibility. A second possibility is that the civilisations of Central America share a common origin with those of Europe and Africa, perhaps a lost continent of Atlantis. A third possibility is that there is a universal bank of knowledge, existing on a different plane from ordinary life, into which shamans or 'wise-men' are able to tap. As shamanism is a worldwide phenomenon, this could explain similarities in beliefs in places widely separated by time and space. I intend to investigate all of these possibilities in a sober and responsible manner. However, the climax of the book will be the suggestion that the Mayan calendar prophesies some sort of cosmic visitation in 2012.

In *The Mayan Prophecies*, I wrote about the 'star-gates'. These are the two points in the sky where the ecliptic[2] crosses the median plane of our galaxy, the Milky Way.[3] In the ancient world, in Europe and Africa, there was a widely held belief that these points were portals through which souls entered or left our world. However, as *The Mayan Prophecies* was primarily concerned with theories concerning Mayan calendrics in relation to sunspot cycles, I only mentioned these in passing. In the present book, I want to go back into this subject in more depth. Drawing on the most recent scholarship, I have discovered that the Maya knew about the star-gates and indeed held many other cosmological ideas in common with Europeans. In particular, they, like the Romans and Greeks, had beliefs about the importance of star-gates as portals to 'higher worlds'. I now intend to explore and elaborate upon these ideas. I will show how they indicate a belief, in the New World as in the Old, that ages are terminated and initiated by the 'gods' at times when these portals are symbolically open.

All this will link in with another of my books, *Signs in the Sky*,[4] in which I presented evidence for believing that biblical prophecies concerning the end of the age are being fulfilled in our own times, with the Sun now positioning itself at a star-gate on the summer solstice. Based on further in-depth research, I will now explain why I believe that the Mayan end-date of 22 December 2012 concerns the other star-gate. It seems to me that the 'opening' of the first star-gate in June 2000 symbolised the start of a process of change that will go on until 2012, when the second star-gate is also symbolically opened. This means that we are currently living through a period of transition. What will happen when this period is ended is a matter of speculation. However, I shall explain why I am of the opinion that the initiation of world ages is indeed, as the Maya believed, connected with long-period, astronomical time-cycles.

Prologue

In 1995, I co-authored a book called *The Mayan Prophecies*.[1]
It was a great success and has since been translated into
more than a dozen foreign languages. At the core of the
book was a theory, first proposed by my co-author, Maurice
Cotterell, that the sophisticated 'long-count' calendar of the
Mayan Indians was based on knowledge concerning
sunspot cycles. Yet how the Maya came by this knowledge
was a mystery then and is still a mystery now. Seeking an
answer to this question, I went on to examine evidence for
contacts between the Old and New worlds in classical times.
I also investigated the possibility that the Maya could have
inherited their knowledge from the lost civilisation of
Atlantis. However, although the book addressed these issues,
it raised as many questions as it answered. In this sense, it
was an unfinished work.

Much has happened in the ten years since it was written,
not least the launching of the SOHO (Solar and Heliospheric
Observatory) satellite in December 1995. This satellite,
which is in a fixed orbit around the Sun, constantly sends
back data concerning solar activity. It has confirmed that
the Sun is much less predictable in its behaviour than was
previously thought. It has shown that the Sun is prone to
huge storms that have a dramatic effect on the Earth's

weather. Not only that, it is now clear that the cycle of sunspot activity is not as easy to predict as was once thought, which is important as this too affects our weather.

People have, of course, been observing sunspots for centuries, ever since the invention of the telescope by Galileo. Whilst their cause is a matter of opinion, it has long been recognised that there is a roughly 11.1-year cycle between successive sunspot maxima and minima. However, we now know that this cycle is at best only part of the story. While it may be true that for most of the past 400 years the Sun was following an 11.1-year cycle of activity, today this no longer seems to be the case. SOHO revealed that in the year 2003, when the Sun should have been moving into a more dormant phase, there was instead a massive increase in sunspot activity. Not only were there many more sunspots than expected, but they were also very active in producing solar flares. This activity reached its peak in October when an unexpectedly powerful Coronal Mass Ejection (CME) erupted from a sunspot and some nine hours later hit the Earth's magnetosphere. This flare produced 'Northern Lights' that could be seen as far south as Texas. Events like these, which seem to be becoming more common, indicate that the Sun's magnetic field is going through major changes.

Any changes to the Sun's magnetic field affects the entire solar system, not least the Earth. This in itself is not too surprising. What is, though, is that the Maya, a native American people whose civilisation was basically Stone Age, seem to have known about 'long-period' cycles of solar activity. Their calendar predicted that the present 'Age of the Jaguar' would come to its end in 2012. We moved into the last *katun* of this age on 15 April 1993.[2] We are, therefore, from the point of view of the Mayan calendar, living through the 'end times'.

New discoveries concerning sunspot cycles are not the only development affecting this book: Mayan studies have also come on apace. Although as early as 1952 Yuri Knorosov, a Russian scholar of great brilliance, published an article on how Mayan hieroglyphs might be deciphered, because of academic intransigence this work was held back for two decades or more. In 1973, the first *Mesa Redondo*, or

Round Table, conference was held at Palenque in Mexico. This brought together many of the world's experts on the art, epigraphy and archaeology of this most exquisite of Classic Mayan cities, with the aim that through concerted effort they might succeed where others had failed in decoding the many Palenque texts. Since then, the world of Mayanology has been in a ferment, as more and more hieroglyphic texts – on stelae (commemorative or funerary stones), door lintels, stucco panels, bark-books, ceramics and even jewellery – have been deciphered. As a result, old ideas of the ancient Maya elite as a pacifistic, other-worldly priesthood have been overturned. It is now known that for much, if not most, of the period of Mayan civilisation, the mini-states that made up their world were at war with one another. In this sense, they were not much different from the city states of classical Greece, whose interminable squabbles eventually led to their conquest, one and all, by the semi-barbarian Macedonians. But like the Greeks, too, the Mayans of the Classic era (*c*. AD 250–1000) left behind them a legacy of majestic buildings, a written script and the enigma of how these things had been achieved.

Although the work of translating and interpreting Mayan hieroglyphic texts was well under way at the time we were writing *The Mayan Prophecies*, as this was still a rather esoteric field monopolised by a small number of experts, I felt it best to concentrate our attention on the more directly relevant – relevant from our point of view – numerical hieroglyphs. This did not actually seem much of a hardship at the time, as our principal interest, where hieroglyphic texts were concerned, was calendric. As the system of decoding Mayan dates has been known for the best part of a century, we did not need to reference more modern research on hieroglyphs. In any case, I felt that to do so would be to complicate matters unnecessarily, as our intention was to show – which we did quite convincingly – how the Mayan long-count calendar fitted with Cotterell's ideas on sunspot cycles.

Unfortunately, there were also undoubted weaknesses in our approach. In retrospect, the work of the Mesa Redondo scholars, particularly the late Linda Schele, would have helped greatly in interpreting the true meaning of the 'Lid of

Palenque': the world-famous covering for the sarcophagus of Pacal the Great. As it is, we presented Maurice Cotterell's personal interpretation for the designs on this extraordinary work of art: an interpretation that I now believe to be wrong in almost every respect. In this new work, I am presenting what I believe is a better interpretation for the meaning behind the designs on the Lid and one which is more in harmony with expert opinion.

That said, there was a great deal of valuable information contained in the old book that would be hard, if not impossible, for readers to find elsewhere. One reason for this is that Mayanology, perhaps more than most branches of archaeology, is restrained by political considerations. Probably the most insidious of these is an academic taboo, which extends throughout Latin America and till recently even in the United States, against discussing the possibility that, prior to the arrival of Columbus in 1492, the New World may have been colonised (or even visited) by anyone other than 'Native Americans'. Of course, 'Native Americans' (for reasons of political correctness, we are no longer allowed to use the term 'American Indians') had to come from somewhere. The assumption is that they were all, diversity of tribes, languages and cultures notwithstanding, descendants of a small group of Asiatic people who entered Alaska across the Bering Straits during the last ice age. It is well known in academic circles (though not publicly admitted) that any archaeologist who publishes a paper claiming that Mexico was visited by Europeans prior to the arrival of the conquistadores is likely to have their digging licence revoked. To even discuss the possibility of ancient mariners crossing the Atlantic is deemed to be a slur on the dignity and reputation of Columbus. Worse still, it implies that the ancient Central American civilisation was not an entirely home-grown phenomenon but owed much to the importation of ideas from elsewhere. The taboo against saying this holds despite the discovery in the Americas of Roman artefacts, Scandinavian runes, Egyptian hieroglyphs and other epigraphic writings that look very much like late Carthaginian script. In actuality, the 'Alaska first' theory – in truly scientific terms it is only a theory – is not even borne out by the archaeological evidence of human settlement.

The taboo against discussing transatlantic contacts once extended to the whole of the North American continent. However, in the United States and Canada, while there is still denial of the possibility of any Roman or Carthaginian links, there is now acceptance that Viking explorers probably did make the voyage to America – as indeed their sagas claim – maybe 500 years before Columbus. Furthermore, very recent work at many sites in both North and South America is undermining the assumption that all native Americans prior to the arrival of Columbus are descended from Alaskan immigrants who crossed the Bering Straits. Archaeo-biological evidence now indicates that South America was settled prior to North America. Recent work at Pedra Furada in Brazil has revealed human occupation there at least 56,000 years ago. This is the earliest known example of human occupation in the Americas and predates by tens of thousands of years the epoch in which Native Americans are supposed to have entered Alaska from Siberia. As there was no land-bridge to South America available, even in the Ice Age, this continent's earliest settlers must have come by sea. If that is so, then there is no reason to believe that North America was not, in the main, colonised in the same way: transoceanic contact between the ancient Old and New worlds has to be a real possibility.

The subject of transatlantic contacts is one we broached in *The Mayan Prophecies*. I am going into it in more detail in this new book. For, regardless of any academic taboos and independently of DNA evidence, there are good reasons for believing that over the millennia the Americas have been visited repeatedly by people from not only Europe but also Africa and even China. Besides evidence such as actual artefacts, there is a congruence of technological ideas and religious concepts. That scholars are in denial of transatlantic contact is all the stranger in that the traditions of many native tribes describe civilisation being brought to them by a bearded white man, known variously as *Quetzalcoatl* (Aztec), *Kukulcan* (Mayan) or *Viracocha* (South American).

Another taboo subject for archaeologists and historians alike is Atlantis. Like the concept of transatlantic contact,

Map of Central America

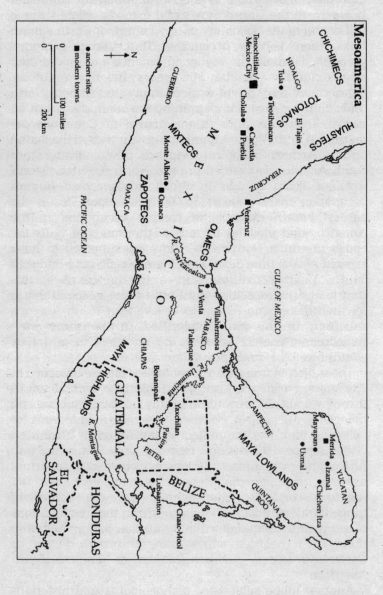

Mesoamerica

CHICHIMECS

HUASTECS

TOTONACS

Tenochtitlan/
Mexico City ■
Tula ●   HIDALGO
    ● Teotihuacan
Cholula ●   El Tajín ●
         Cacaxtla ●
         Puebla ■   VERACRUZ

GUERRERO

MIXTECS   M
Monte Albán ■   E
ZAPOTECS   X
Oaxaca ■   I
                   C
PACIFIC OCEAN   OLMECS   O
              R. Coatzacoalcos   Veracruz ■

              La Venta ●   GULF OF MEXICO
              Villahermosa ■
              TABASCO
              Palenque ●
              CHIAPAS R. Usumacinta   CAMPECHE
              Bonampak ●
MAYA   Yaxchilán ●   Mayapán ●
HIGHLANDS   R. Pasión   Mérida ■
GUATEMALA   PETEN   MAYA   Izamal ●
              R. Motagua   LOWLANDS   Chichén Itzá ●
EL   Uxmal ●   YUCATAN
SALVADOR   Lubaantún ●   BELIZE
HONDURAS   Chaac-Mool ●   QUINTANA
                                            ROO

N

0
0   100 miles
     200 km

● ancient sites
■ modern towns

for some reason the idea of a lost continent also provokes apoplexy among the academic community. The oldest reference to this legendary island is in the Greek philosopher Plato's dialogue *Critias*. He states categorically that Atlantis lay beyond the Pillars of Hercules (Gibraltar) and was not to be confused with the 'real' continent that lay beyond it on the other side of the Atlantic Ocean. Unfortunately, Plato's words have not only been ignored by academics but have been misquoted by many alternative writers too. Instead of accepting his geography – that Atlantis lay on the other side of the Atlantic Ocean – the lost continent has been relocated everywhere from the British Isles to the Eastern Mediterranean to Antarctica. Yet Edgar Cayce, probably the greatest psychic of the twentieth century, was unequivocal that not only did Atlantis exist, but it was located in the only place it could have been based on the depth of the ocean in the region described by Plato: the Bahamas. This again is a subject that was raised in *The Mayan Prophecies* but is discussed in greater depth here. I have endeavoured to show that not only did Atlantis exist – and go under the waves – but it was the root from which all the later Mesoamerican civilisations sprang.

Finally, there is the prophecy itself. If the present age is scheduled to end in 2012, what might this mean for us? Should we be worried or will this date turn out to be of no greater significance than the millennium year of 2000? This is a subject I have addressed, I hope, in a comprehensive way. I am able to show that this is not an arbitrary date but one closely linked to observable astronomy. That being so, we have to ask ourselves just how the Maya, a primitive Stone Age people even if possessed of some of the trappings of civilisation, could have known thousands of years in advance and to the day just how the sky would be on 21–22 December 2012. This is one of the world's great mysteries. I hope that here I have provided a thesis that may go some way towards answering this and other important questions.

CHAPTER 1

Mexico Revisited

The plane circled slowly, making ready to land. In the distance was 'Smoking Mountain': *Popocatepetl* in the Aztec language but affectionately known as Popo. Though relatively quiescent on our approach, a plume of smoke rose from it, like breath emanating from the nostril of a sleeping dragon. Even without this telltale sign, I could see that this was not an ordinary mountain peak. Its sharply pointed profile and the treacle-like rock-flows down its sides betrayed its nature. They testified to a long history of terrifying eruptions, some of which, we know, were witnessed by the Aztecs. Today, all was quiet. The peak of the volcano, the second highest in Mexico,[1] glistened above a tablecloth of clouds in the morning sunshine. Its cone was fringed with snow, which seemed odd given its location in the tropics and its proximity to deserts and jungles. The snow, though, was all down to Popo's colossal height. At 16,400 ft, it is twice the height of Mount St Helens in Washington State and one and a half times that of Mount Etna, Europe's largest and most active volcano. Little wonder, then, that the Aztecs were in fearful awe of Popo's power.

It was mid-March 1998, nearly three years after the publication of *The Mayan Prophecies*, and I was returning to Mexico. Unlike my last trip, when I was able to travel incognito, this time I was not alone but at the head of a tour group. I felt anxious as our plane touched down, not because I feared a crash or that Popo might blow his top while I was too near for comfort. No. I was afraid that the tour, organised by a now defunct magazine called *Quest*, would be a failure. Disembarking from the plane and feeling the warm embrace of Mexico City, those fears melted away. I felt sure that it would be hard for our guests, most of whom had never visited Mexico before, not to enjoy themselves in such a vibrant country.

Even so, I still had some anxieties on my own account. Since writing *Prophecies*, I had made certain further discoveries concerning the Maya and their extraordinary calendrical knowledge. I had also made some attempt at learning the rudiments of their hieroglyphic language which, it is now clear, was on a par with that of ancient Egypt. I was particularly looking forward to taking another look at some Mayan temples, not because I expected to be able to read the hieroglyphs inscribed on many of them – my knowledge didn't extend that far – but rather to 'feel' them. For it is my experience that every place has its own *spiritus loci*, or atmosphere, which although intangible is just as real as odour and appearance. I was anxious to see if, when revisiting places I had been to before, they evoked the same feeling: that the pre-Columbian civilisation of Central America stretched back much further in time than is generally believed. I also wanted to 'tune in' on the sites I was visiting, to allow my unconscious mind to speak to me. This is a technique I have used many times and in places as diverse as the pyramids of Egypt, the Stone Age monuments of Britain and the wilderness of Australia. I have found through experience that if I give my subconscious mind free rein, it has the ability to present me with the answers to questions that my conscious mind cannot figure out. In ways that are quite inexplicable in ordinary terms, the answers come to me in the form of unexpected meetings, accidentally finding a particular page I had previously overlooked in a book or simply noticing something when walking around. As we got off the plane, I was in this open-minded state, aware that there were even deeper levels of mystery concerning the Mayan calendar than Maurice

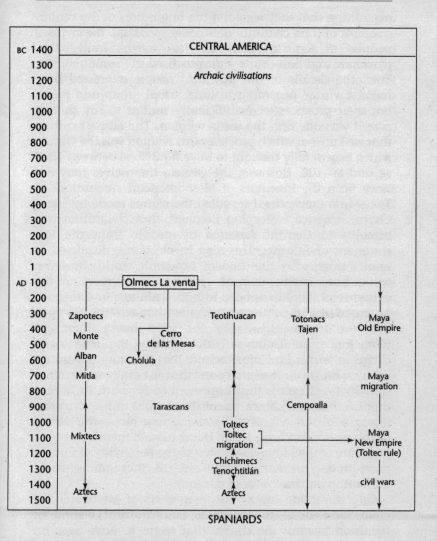

Figure 1: Chronology of Central America

Cotterell and I had explored in *The Mayan Prophecies*. I was hoping to receive further clues as to what this mystery might be and how it fitted with the broader picture of world civilisation.

Pick up any book on the Maya and it will begin by

explaining that they were not the originators of the central concepts of their civilisation. Broadly speaking, the civilised peoples of Central America – those who built cities, pyramids and ball-courts – all practised the same religion. True, the details of the way that people expressed their religion varied according to time, tribal group and place, but their practices were sufficiently similar to say that at core it was one and the same religion. The earliest culture that we know of which practised this religion was the Olmec, which is generally thought to have flourished between 1200 BC and AD 100. However, the Olmecs themselves may not have been the inventors of Mesoamerican civilisation. In *The Mayan Prophecies*, I repeated the claims made by Edgar Cayce, America's 'Sleeping Prophet', that civilisation was brought to Central America by people from the lost continent of Atlantis. This idea is, of course, dismissed as total nonsense by the modern academic world. However, while hard archaeological evidence in support of the existence of Atlantis is hard to find – still less in the epoch proposed by Plato of the tenth millennium BC – the theory of the lost continent is very far from being disproved altogether. Careful study of Plato's writings, the *Timaeus* and *Critias* in particular, convinces me that the story of Atlantis is based on truth. It seems to me that his dialogues contain a memory of events that happened so far back in history that all that was known about them, even in his own time (some two and a half millennia before our own), was a distant echo. As I have found traces of such memories in the literature of the Maya, I was alert to the possibility of finding more among the detritus of the ancient cities that we would be visiting on the course of our tour.

One stumbling block that researchers of Mesoamerican civilisation have to overcome is a natural feeling of revulsion towards the cruelty that seems to have been an essential part of the indigenous religions. This cruelty is reflected in the grotesque nature of so much Mesoamerican art and even more so in the preserved literature. There is no doubting that, as witnessed by the conquistadores, the pre-Christian religion of Central America was frightening. The Aztecs in particular practised human sacrifice on a massive scale. Superstitious to an amazing degree, they would tear out the still-beating hearts of sacrificial victims and offer these as food to the Sun. As if this were not bad enough, they would on other occasions flay their victims while still alive and then wear their skins for rituals that could last for weeks.

Unfortunately, images of religious practices involving human sacrifice and what to us is unspeakable cruelty are what first confront the would-be investigator of the Maya as well as the Aztecs. A day trip to the largest of all their cities, Chichen Itza, is enough to confirm that not all the Maya were the peace-loving sages of romantic myth. Like Solon's Greeks, they engaged in inter-city warfare on a routine basis and frequently tortured or beheaded captives. If this were the be-all and end-all of Mayan civilisation, then my gut instinct would be to have nothing to do with it. However, it is not all: there was clearly something else going on. Their cities exude a particular feeling or atmosphere. That feeling intimates that there is a great mystery masked by the gory details of ritualistic slaughter. One comes away with the indelible impression that what the early Spaniards encountered (and what is portrayed on thousands of murals and pieces of sculpture) is but a decadent remnant of something much higher. It is that 'something' which I found interesting on my first trip to Mexico and it has been drawing me back ever since. However, the principal reason I find the Mayan civilisation interesting is because I believe it has a message for our own times. To find this message and to interpret it in terms that we can understand today is, for me, reason enough to overcome feelings of initial repulsion towards much of their art. It was why I was returning to Mexico once more to examine *in situ* the art and records of this past civilisation.

Arriving in Mexico City, I was struck by a sense of calm. There, in defiance of all expectations to the contrary, some twenty-six million people were still managing to live more or less in harmony with one another. Twenty-six million, which is more than the combined total population of London and New York, is for me an unimaginable number of human beings. That so many souls are somehow managing to eke out a living in such a relatively small area seems nothing short of miraculous. Even so, there is no doubting that Mexico City is under pressure. Regardless of the considerable wealth generated by the booming oil industry, continued immigration from the rest of what is mostly a very poor country condemns it to third-world status. The true situation is most obvious from the large number of squatter camps which, like toadstools round a great oak, form a 'fairy ring' around the periphery of the city itself. Despite these added burdens, Mexico City maintains its dignity. Somehow, despite all the pressures from over-population, the central area was relatively clean. Indeed, even the famous atmospheric pollution, reminiscent of the London smogs of my childhood before the Clean Air Act of 1956 made these a thing of the past, was not present. Driving along the Avenida Paseo de la Reforma, the main thoroughfare along which many foreign banks and international companies have their offices, the air seemed, if anything, cleaner than it was the last time I was in Los Angeles. Far from being blocked out by smog, the morning light was intense. It made the glass buildings on either side of the road sparkle like a forest of giant crystals set out in a park. In this part of the city, at least, there was the sense of a country ready and eager to take its place at the top table of capitalist nations.

The *Zocalo*, or town square, lies at the very heart of the old city. It is overlooked on the western side by the enormous bulk of Mexico City's baroque cathedral, while the northern is dominated by the National Palace. This majestic building, less ornate but considerably more attractive than the cathedral, was built out of the remnants of Aztec pyramids. The official residence of the president, it is also a major tourist destination. The principal attraction, however, is not the building itself but rather the series of murals painted by Mexico's most famous artist: Diego Rivera. In graphic detail, these tell the country's story, from the splendour of the pre-colonial civilisations, through the days of colonisation by Spain, to the dark days of chaos that followed the execution of the emperor Maximilian, to the revolution that led to the present-day republic.

Politically correct from the standpoint of a card-carrying communist like Rivera, the murals depict in morbid detail the atrocities committed by the Spanish conquistadores while glossing over the barbaric cruelty of the Aztec Empire that preceded them. However, we cannot blame Rivera for his selective memory. Mexico is a city and country that has struggled for centuries to come to terms with its history: one where human life was once held as cheap as that of chickens. Worship of the gods involved human sacrifice on a scale now barely imaginable. Mexico's march into the modern era, for better or worse, began in ways that even a superstitious people such as the Aztecs could not have predicted: invasion from across the Atlantic. Their overthrow was to be one of the great turning points of history, the repercussions of which are still with us to this day.

THE CONQUEST OF MEXICO

The first of several Spanish expeditions to mainland America occurred in 1511. Leaving the relative safety of Cuba, a small ship sailed out from Havana and was wrecked off the coast of the Yucatán Peninsula, on the Island of Cozumel. The Spaniards came ashore looking for gold but instead found a land populated by savage tribes, most of whom were not at all pleased to see them. Nearly all of the Spaniards were killed: in battle, through disease or as unwilling offerings to the gods. At the time, it must have seemed an inauspicious start. Further reconnaissance missions to the Yucatán itself fared little better. The Spanish quickly learnt that though there had once been great cities in the area, these had nearly all fallen into ruins and the native people had reverted to almost subsistence living. Of gold, there was very little to be found, and even the land seemed poor in comparison with Cuba and the other Caribbean islands. Not surprisingly, for the time being they turned their backs on the Yucatán and started exploring elsewhere.

In 1519, Hernán Cortés led a much larger and better armed expedition, comprising a fleet of eleven ships, into Cozumel. His too was a short stay but long enough to impress on the natives that this time the Spanish meant business. A trip along the coast of Yucatán quickly revealed that his fortune was not to be made in this poverty stricken land. Accordingly, he turned his attentions north-westwards to what he reckoned to be a much better proposition: the Aztec Empire. Fate was to prove him right.

One fortunate consequence of his short stay in Cozumel was the rescue of a captive Spaniard named Geronimo de Aguilar. He was one of only two survivors from the 1511 expedition and had by now learnt the local language. As dialects of this same Mayan language were spoken throughout the Yucatán and even along the Gulf coast of Tabasco, Aguilar was to prove invaluable as a translator. On Good Friday of that year, Cortés arrived on the coast of Mexico itself, at the mouth of the Rio de Grijalva. His force consisted of 508 soldiers, about 100 mariners and 16 horses. Among their numbers were 32 crossbowmen, 13 musketeers and crews to maintain and operate several brass field-guns. Thus armed, and with plenty of gunpowder and shot, they set out on one of the most amazing expeditions in history: the conquest of Mexico.

As before in the Yucatán, the Spaniards initially met with stiff resistance. However, it did not take them long to impress upon the local natives that not only was battle futile but an alliance might prove fruitful. Hearing of the enormous wealth of the Aztecs and how greatly they were feared by the coastal tribes, Cortés promised the latter that, if they would make peace with him, he would end their servitude. A treaty was agreed and several women slaves were delivered to Cortés as a token of this new friendship. Among these was a remarkable lady whom the Spanish named Dona Marina. Young, intelligent and good-looking, Cortés would later take her as his mistress. More importantly, she was a good linguist. Already able to speak both Yucatecan Mayan and the Aztec language of Nahuatl, she very quickly learned Spanish. Her services to Cortés as a translator were to prove even more useful than Aguilar's.

Having thus demonstrated Spanish military superiority to the local people, Cortés extended the remit of his mission from trading to colonising. Overriding opposition from some of his captains, he founded the city of Vera Cruz and persuaded his soldiers to elect him to the rank of captain-general. He then burnt all but one of his ships, this being despatched to the king of Spain with a sizeable treasure that included valuable gifts from the Aztec emperor Montezuma.

With friendly relations established on the coast, he felt able to march inland, leaving only a small garrison of older men to hold on to Vera Cruz.

THE EMPIRE OF THE AZTECS

The Aztecs were a nation of warriors who only arrived in the valley of Mexico around 1325. They were therefore themselves still relative newcomers when Cortés arrived on the scene in 1520. Tradition states that their original homeland was a place called Aztlan, which, because of the similarity of language between the Nahuatl-speaking tribes of the Mexico City area and others, such as the *Yaqui*, who live on the border lands of Mexico and Arizona, is assumed to have been well to the north of Mexico City. However, Aztec traditions also say that Aztlan was an island city. This, obviously, could not be the case if it was located in northern Mexico: a region not of water but desert. It is also well to note that the 'Aztecs' did not call themselves by this name. Whatever their true origins, they were known to themselves and their neighbours as the *Culhua-Mexica*.

According to legends which they related to early Spanish chroniclers, they were led to the Valley of Lake Texcoco, where Mexico City now stands, by a seer called Tenoch. He was told in a dream that his people must continue wandering until they came to a place where they would see an eagle fighting a serpent. At the time of their arrival, the Valley surrounding Lake Texcoco was occupied by five other tribes. They, naturally, were not at all keen to share the land with the newcomers but offered them instead an island in the middle of the lake. This island was uninhabited, as it was the home of numerous poisonous snakes. It was no doubt hoped by the indigenous tribes that, should the Mexica decide to take them up on the offer and stay in the neighbourhood, the snakes would deal with them. If this was their expectation, they were to be sorely disappointed. Going out to the island, Tenoch and his followers saw the sign they had been told to look out for: an eagle grappling with a serpent. Delighted by the fulfilment of the prophecy, they accepted the island as their new home and set about building the city of Tenochtitlan, so-named after its founder.

As far as they were concerned, snake meat was a delicacy. Rather than being put off by their presence on the island, they regarded these as a further gift from the gods.

Within a very short time, the Mexica became the dominant tribe of the region, uniting with the surrounding tribes to form a powerful nation. The belief systems of this nation, though partly imported with the Mexica, were largely indigenous. In the main, their religion was derived from an earlier people they called the Toltecs, who had once ruled most of Mexico from a fabled city called Tula. The true location of Tula has been a matter of academic debate for centuries but till recently was assumed to be the city of Tula, about 50 miles north-west of Mexico City in the province of Hidalgo. At first sight, this seems a reasonable suggestion, if only because of the name. However, this city is not particularly old, meeting its end in AD 1156. More recently, it has been realised that *Tollan* means 'place of cattail reeds' and was used generically as a designation for any sacred city built in imitation of an original, archetypal Tollan. Thus, the 'Tula' in Hidalgo appears to be just one such imitation. A much older and more authentic site for the original Tula of the Toltecs is the ruined city of Teotihuacan. More will be said about this later.

The Aztec was the most bloodthirsty and frightening of all the empires the conquistadores encountered in either North or South America. They practised human sacrifice to an alarming degree, the removal of the still-beating hearts from living victims being the central mystery of their religion. For the consecration of one temple alone, it is estimated that some 20,000 captives were sacrificed. Unfortunately, this was not an isolated incident of what we would consider utterly psychopathic behaviour. Each year, they would kill some 50,000 victims, either offering their hearts to their idols or, perhaps even more macabre, flaying their victims alive and themselves wearing the flayed skin until it became so putrid it began to fall apart.

The reason they practised human sacrifice was that they believed their gods needed to be fed. Their favourite food was evidently human life force, most visibly seen in the twitching hearts presented to them on the banqueting tables of the temples. To satisfy their voracious appetite for this delicacy, there was a whole warrior caste, organised into regiments. It was their job to fight wars against neighbouring tribes, not so much for material plunder but to capture living victims. To this end, the Aztecs encouraged insurrection among their subject peoples. This gave them a pretext for sending in their army to take away prisoners. Naturally enough, this behaviour won them few friends and they were greatly feared. The one hope that their many enemies cherished was a prophecy that one day Quetzalcoatl, a bearded white man whose Toltec kingdom had been usurped centuries earlier, would return. It was expected that he would one day reclaim his kingdom from its usurpers. Sword in hand, he would once and forever end the Aztec dominance. His just rule would inaugurate a new age of peace, prosperity and justice. According to this prophecy, his return would take place on a year named 1 Reed on the Aztec calendar. One can imagine, therefore, with what trepidation the Aztec ruler, Montezuma II, heard news, in that very year, that bearded white men had landed on the coast of Mexico. When the Aztec ruler first met Cortés, he assumed he was indeed the returning Quetzalcoatl. By the time he realised this was a mistake, it was too late. In a campaign lasting little more than two years, the Spaniards took control of Mexico and set about converting it to Christianity. What had been the most powerful and richest empire of the Americas was turned into a colonial province of distant Spain.

Cortés's first impressions of the Aztec capital Tenochtitlan were mixed. He was impressed by its size, its opulence, its advanced architecture and the Aztecs' clever use of artificial lake-islands for raising crops. However, he was appalled by the stench of death that hung over this otherwise civilised community. He quickly discovered that this emanated from the temples on top of pyramids caked with the dry blood of sacrificial victims. The Aztecs' callous indifference to suffering and their idolatrous rituals filled the Spanish with abhorrence. Cortés, appalled by both the stench and what he saw, sought to banish the evil spirits with the sign of the cross. In a flagrant breach of protocol, he insulted the Aztec gods, saying:

'Señor Montezuma, I do not understand how such a great prince and wise man as you are has not come to the conclusion, in your mind, that these idols of yours are not gods, but evil things called devils, and so you may know it and all your priests see it clearly, do me the favour to approve of my placing a cross here on the top of this tower, and that in one part of these oratories where your Huichilobos and Tezcatlipuca[2] stand we may divide off a space where we can set up an image of Our Lady [an image which Montezuma had already seen] and you will see by the fear in which these idols hold it that they are deceiving you.'[3]

Montezuma's angry reply to Cortés's rather insolent suggestion reveals the extent of the divide that existed between their two cultures:

'Señor Malinche,[4] if I had known that you would have said such defamatory things, I would not have shown you my gods. We consider them to be very good, for they give us health and rains and good seed times and seasons and as many victories as we desire, and we are obliged to worship them and make sacrifices, and I pray you not to say another word to their dishonour.'[5]

This exchange, which took place at the top of the then most important pyramid of Mexico, marks the turning point of the relationship between Aztecs and conquistadores. Now that religion was involved, there could be no compromising on either side. Though Montezuma was undoubtedly hurt by the slight on his gods, Cortés and the Spanish too hardened their hearts. They determined that the Indians were devil-worshippers who, for their own good and forcibly if necessary, needed to be converted to Christianity. Although they suffered some setbacks, in the end it took the Spanish only two years to defeat the Aztecs and impose Spanish rule throughout their empire. They destroyed the Aztec pyramid of Tenochtitlan with gunpowder and recycled its stones for the construction of new, Spanish-style buildings. Any idols that could be found were also either destroyed or buried, the intention being to erase all memory of the Aztec religion.

Once they had conquered Tenochtitlan and taken the heart of the Aztec Empire, it took only a few more years for Cortés and his men to subdue the rest of Mexico. In the main, they found the subordinate tribes only too willing to be liberated from their former masters and to accept the conquistadores as their new overlords.

Having dealt with the Aztecs, the Spaniards now turned their attentions back towards the Maya, whom they had left to their own devices since 1511. In December 1523, Don Pedro de Alvarado began the subjugation of what is now Guatemala. This highland region was home to a number of tribes but most notably the Quiché and their cousins the Caqchiquels. In a campaign of great savagery, Alvarado subjugated the Quiché, massacring their chieftains and burning their capital city of Utatlan. A year later, in 1524, Cortés sent Cristobal de Olid, one of his most trusted lieutenants, to secure the part of the Central American isthmus that now comprises the countries of Honduras and Nicaragua. Olid founded the city of Triunfo de la Cruz, in the Bahia de Tela.[6] However, he did not enjoy his new town for long, as he was assassinated shortly afterwards. It was then left to Cortés himself to secure the authority of Spain in the region, marching down from Mexico the following year to do so. Not far from Olid's foundation, he established a colony of his own that was first called Puerto Caballos and is now known as Puerto Cortés. The remaining Mayan tribes that were not yet in the grip of the conquistadores were now effectively surrounded on all sides.

The Mayans' best defence against Spanish aggression was the impenetrable nature of the forests that grew luxuriantly over most of their lands. North of Puerto Cortés, in the Yucatán Peninsula, most of the country was covered in such forest and conquest was not an easy proposition. Nevertheless, serious attempts at colonisation were made by the Spanish from 1527 onwards. However, until the founding of Merida, in 1540, they were unable to establish a permanent foothold. Even then, perhaps because the country was lacking in obvious resources such as gold and silver, the Yucatán was unattractive to Spanish immigrants.

As a result, the Maya remain to this day more purely aboriginal than any of the other indigenous peoples of Central America. Though forced to convert to Catholicism, they have nevertheless preserved – albeit in modified form – many of the characteristics of their old religion. Their languages too have survived. These comprise at least 30 different dialects, the most important language families being Yucatecan, Cholan and Quichean. This has been extremely fortuitous, as without this survival it would have been impossible to decode the Mayan hieroglyphs: a singular achievement of the late twentieth century.

REPORTS OF THE FRIARS
The conquest of Mexico was in many ways a remarkable achievement and few today would dispute that Cortés was right to put an end to human sacrifice. A more difficult charge – one that is seldom challenged, still less refuted – is that directly or indirectly (through disease and over-work) the Spanish were responsible for the deaths of millions of indigenous Mexicans. It is certainly true that the Spanish, many of whom were little more than pirates, were to blame for many deaths. Yet contemporary accounts reveal that this was not the whole story. Shocked by the profound change to their society, many Indians were unable to cope psychologically with the change. It is as though so profound was their sense of cultural loss that they lost all will to live. They either hanged themselves or simply wasted away.

Why this should be was something that I have long wondered at. The answer to this thorny question became clear only after I wrote *The Mayan Prophecies* and will be explained later on in this book. What cannot be denied is that the conquest paved the way for many undesirables from Spain to migrate to Mexico. These second-wave conquistadores were often even more brutal in their treatment of the *indigenas* than the original conquerors. Anxious to make fortunes out of agriculture and cattle farming, they regarded the natives as little more than a useful source of slave labour. However, not all these new arrivals were slave masters. Some were pious men, anxious to save the natives' souls and to preserve the memory of the best of the pre-Columbian past before all record was gone forever.

The most famous of these, if only because he wrote a number of books, was a Franciscan friar named Bernadino Sahagun. A young man, only just out of college, he sailed to Mexico in 1529 – barely a decade after Cortés's first arrival. On board his ship were several Indians who had been brought to Spain as curiosities and were now returning home. An able linguist as well as a historian, Sahagun used his time while crossing the Atlantic to learn from them the rudiments of Nahuatl, the native language of the Aztecs. Within a very short time, he became so proficient that he was able to speak it nearly as well as a native. This ability to talk to them in their own language, along with his sympathetic spirit, enabled him to win the confidence of the people with whom he came into contact. Consequently, they trusted him with many of their secrets.

His books, in which he gives a lot of detail concerning the Aztec calendar and the festivals held at various times of the year, make disturbing reading. The sections on calendars make it clear that nearly every Aztec feast day involved human sacrifice of not just men but also women and, most distressing of all, children. Given this evidence, it is hard to see how the Spanish authorities could have done anything other than suppress the native religion, so imbued was it with human blood.

Sahagun was not the only Spaniard to elicit secrets from the indigenous people among whom he worked. Another towering figure was Friar Diego de Landa. A member of the Franciscan

Bernadino Sahagun

37

order like Sahagun, he arrived in the Yucatán at some time in the 1540s. Bishop Landa, as he later became, was a man who made many enemies – among the expat Spanish community as well as the indigenous Mayans. Most famously, in 1562, he held an auto-da-fé during which many Mayan bark-books were burnt. Today, these would be considered priceless antiquities, but at the time they were held to contain devil-inspired teachings that could only hold back the successful conversion of the Maya to Christianity.

To a Christian missionary like Landa, it made sense to burn all the books that could be found and thereby root out the source of what he saw as an evil religion. Fortunately for us, he did more than just burn books: he also wrote a history of Yucatán entitled *Relación de las Cosas de Yucatán*. This seminal work contains a condensation of all that he learnt about the indigenous Mayan culture of the region. It contains details about festivals and gods, costumes and calendrics, as well as the more prosaic accounts concerning the flora and fauna of the region. Most important of all (as we will see later), Landa included in his account what has turned out to be a 'Rosetta Stone' for the successful decipherment of Mayan hieroglyphs.

Because this is a highly complex subject in its own right and it was not really relevant to our thesis, I did not discuss the decoding of Mayan hieroglyphs in *The Mayan Prophecies*, other than those which were calendrical in nature.[7] This was possibly a mistake, as over the course of the past 20 years or so, an enormous amount of good work has been done in this field. The work of Mayanologists such as David Freidel, Linda Schele, Peter Matthews, Michael Coe and many others cannot justifiably be ignored if we wish to have a fuller appreciation of Mayan mysteries. As we will see in later chapters, it has led me to a profound reappraisal of the Mayan prophecies and some surprising new discoveries concerning the possible source of Mayan astronomical knowledge.

Others besides the Franciscans took an interest in pre-conquest Mexican culture. In the late seventeenth century, a former Jesuit called Don Carlos de Sigüenza y Gongora succeeded in amassing a large collection of manuscripts

Figure 2: The Aztec calendar stone

and paintings. One of his friends was Don Juan de Alva, who was the son of Alva Cortés Ixtilxochitl, a man directly descended from the kings of Texcoco.[8] Ixtilxochitl was educated in Spanish as well as the Aztec language of Nahuatl and wrote an extensive history of Mexico in Spanish. Sigüenza was a professor of mathematics at the University of Mexico and a keen astronomer. He was therefore naturally intrigued to discover that Ixtilxochitl's book had a great deal to say on the subject of the native calendar system that was in use prior to the Spanish conquest. He discovered that the Aztecs spoke of a calendar stone which disappeared after the conquest.

Apparently, it was used to keep an accurate chronology over long periods of time by reference to cycles of 52 and 104 years.

Using the books in his collection, Sigüenza was able to glean other important information, including such important dates as the start of the Aztec Empire with the founding of Tenochtitlan in 1325. He was also able to set out a rough chronology of earlier empires beginning with the Olmecs, the so-called 'Rubber People', whose civilisation preceded the Toltecs and who lived in the Tabasco region of Mexico, where rubber trees grew. He believed the Olmec civilisation had its origins in the legendary island of Atlantis and that, among other things, they were responsible for building the impressive pyramids at Teotihuacan.

In 1697, Sigüenza received a visit from an Italian named Giovanni Careri, who was in the process of circumnavigating the world. This was still a very rare accomplishment in those days, which perhaps explains why the Spanish authorities allowed him into the country at all. During his time in Mexico City, Sigüenza showed him his archive and explained his calendrical theories to him. When Careri returned home, he wrote about these things in a book called *Giro del Mondo*, which documented his journey round the world. Though his memories of what Sigüenza told him of his discoveries were far from lucid, it is just as well he put them into print. Not long afterwards, the professor died and his library was seized by the Jesuits. Just what became of his papers is a mystery, but it is likely they were destroyed after the Jesuits themselves were expelled from Mexico in 1767.

Although much of what Careri wrote about Mexico was met with derision in Europe, his tale of the calendar stone sparked the interest of at least one other explorer: Baron Alexander von Humboldt. A friend of Goethe, Schiller and Metternich, he was famous throughout Europe as a naturalist. In 1798, he intended to go with Napoleon's *savants* to Egypt but was unable to because the ship he was booked to go on was lost in a storm. Instead, he went westwards, across the Atlantic, in order to explore the flora, fauna and topography of Spanish America. In 1803, he arrived in Acapulco, the main port on the Pacific coast of Mexico, bringing with him a large amount of scientific equipment. Arriving in Mexico City, he too befriended the Spanish authorities and was given access to classified archives. As luck would have it, some 12 years earlier a large carved stone had been found near the cathedral. Humboldt inspected this and agreed with a local historian called Leon y Gama that it had to be the same 'calendar stone' spoken of by Sigüenza. Furthermore, he recognised in some of the 18 symbols used by the Aztecs to depict months close parallels with the East Asian zodiac. Contrary to the expressed opinions of the church, he concluded the stone was not an altar used for sacrifice but had astronomical and calendrical significance.

Following the Mexican Revolution of 1821, it became much easier for non-Spanish Europeans to visit the country. The first of these was William Lubbock, who sailed from Liverpool in 1822. Like von Humboldt, he was much impressed with the calendar stone and took plaster casts of it. When he got back to London, he put these on display, along with much else, in a gallery in Piccadilly. Lubbock's exhibition, like the recent Aztec exhibition at the Royal Academy during the winter of 2002–3, was a sell-out. It established that pre-Columbian Mexico was more than just a land of barbaric superstition. This, however, was just the beginning. Soon, many other Europeans and Americans would discover that the mysteries of Mayan civilisation were worth unravelling.

CHAPTER 2

The City of the Dead

Leaving the airport, the group and I boarded a coach for our hotel, some 40 km from the centre of Mexico City. The hotel was on the campus of what is considered by many to be Mexico's primary archaeological wonder: the lost city of Teotihuacan. I don't say 'lost' because the ruins of this magnificent complex are hidden away in some inaccessible canyon or remote jungle. Since the ruins of Teotihuacan are only a bus ride away from the centre of the world's largest city, that would be the very opposite of the truth. No, what is lost is not the location but the purpose of this great city. Part Rome, part Giza, part Mecca, it seems to have dominated the whole of Central America for at least 1,000 years and probably much longer if the truth be told. Then, with a suddenness that is hard to explain, it was abandoned by its residents. Its major buildings were deliberately burnt and left to the local scrub. This could be understandable were there signs that the city had been attacked and sacked by some invading force. Yet this does not seem to have been the case. Archaeologists working in the field are adamant that the destruction of the city was done by the Teotihuacanos themselves. This is not normal behaviour by anyone's standards. Yet it does appear that whatever outside influences might have been involved, they were peripheral to the event and not its cause.

The ruins of Teotihuacan were sacred to the Aztecs long before the arrival of the Spanish. It was they who called it Teo-ti-huacan, meaning 'Sacred City of the Gods'. They called it by this name because it featured prominently in their myths concerning the creation of the fifth sun: that of our present age. According to this myth, at the end of the previous age survivors of a catastrophe gathered at Teotihuacan. The gods set about creating a new race out of this remnant.

At this time, the Earth was shrouded in darkness, for there was no sun. Accordingly, they called for a volunteer to become the new sun, whose light would bring the dawn. An arrogant god called Tecuciztecatl volunteered for the honour of becoming the new sun, but the others voted for a sickly old god called Nanahuatzin. The two gods then spent a period of time fasting, while the other gods prepared a sacrificial pyre. Tecuciztecatl put on his best clothes and was richly adorned with quetzal feathers, whereas his rival could only afford a suit of paper. Nevertheless, it was Nanahuatzin who was the first to be brave enough to jump into the fire, being immediately immolated. After this humiliation, Tecuciztecatl quickly followed his example.

The two gods having sacrificed themselves, all eyes turned anxiously to the horizon to see Nanahuatzin resurrected first as Tonatiuh, the sun god, to be followed by Tecuciztecatl, the new god of the moon. However, the sun and moon, though now visible on the horizon, refused to move unless they were fed. To satisfy their hunger and induce movement, Quetzalcoatl, the Plumed Serpent (sometimes identified with the planet Venus), sacrificed the remaining gods by removing their hearts. Naturally, this selfless gift of the gods required a reciprocity from mankind. Accordingly, so that the sun would continue to rise each day, it became a requirement for future generations to match the gods' generosity and to offer it human hearts. According to this myth, it was over the sites where Nanahuatzin and Tecuciztecatl immolated themselves that people later built the Pyramids of the Sun and Moon. These became the core of the city of Teotihuacan, which grew to be the largest in the whole of the Americas. Though the city was abandoned at some time around AD 750, the site was greatly revered by the much later Aztecs. Once a year, they would gather there to carry out blood sacrifices of their own in honour of the earlier sacrifice of the gods.

Rather as Stonehenge excites us all the more because it is in a state of ruin and we can't be absolutely sure of its purpose, so too were the Aztecs amazed by the remains of Teotihuacan, deserted long before they arrived in the valley of Mexico. The reason why the Teotihuacanos abandoned their city is one of the great mysteries of history. It was discussed in Erich von Däniken's famous, if at times outlandish, book *Chariots of the Gods?*, first published in 1967. Von Däniken's take on the situation was unorthodox, yet simple: the people of Teotihuacan, as well as the former inhabitants of other cities in Mesoamerica that were abandoned at around the same time, had simply been 'taken'. Like the heroes of Steven Spielberg's movie *Close Encounters of the Third Kind*, they had willingly joined 'the gods' – extraterrestrial spacemen – leaving our Earth for new homes among the stars.

While this idea had (and has) considerable appeal for many people, it seems to me to be at best naive. There are, for one thing, problems of logistics to be considered. Would the aliens, kindly as they might be, understand the basic needs of humans for air, food, drink, toileting facilities, bedding and recreation? How would they protect their live cargo from the deadly effects of cosmic radiation and the wasting of muscle and bone that would inevitably occur on any long space journey? Then, assuming that they arrived safely on some distant planet that was sufficiently similar to our own to support our sort of life, how would terrestrial humans be protected from the myriad of bacteria? We know from history that when Hernán Cortés and his men arrived in Mexico, they brought with them the smallpox virus that decimated the native population. Conversely, Christopher Columbus and his crew are blamed for bringing syphilis back to Europe. If such unforeseen and unwanted epidemics can be caused by people migrating within the relatively closed world of planet Earth, how much greater must be the likelihood of this happening if we should come into contact with people similar to us but from other worlds? These are questions which have occupied me for decades. Unless one is willing to entertain self-deception on a mammoth scale, it does not at present seem possible to escape the conclusion that we, in our bodies, are not able to live anywhere else but here on Earth. Conversely, extraterrestrial aliens, assuming they are anything like us, would have similar medical difficulties in adapting to conditions here on Earth. That is not to say that physical contact with alien species is impossible, but simply that it would be highly dangerous for all concerned.

Since my last visit to Teotihuacan, I had read much about Mesoamerican civilisation and begun to realise that it did have an extraterrestrial dimension, even if this did not stretch to the sort of physical contacts described by von Däniken. Astronomy was undoubtedly an integral part of the religion of Central America. However, Mesoamerican astronomy was not the rational science that we know and understand today but something more akin to astrology. The Teotihuacanos, like their successors the Aztecs, lived in a world circumscribed by ritual and captive to superstition. Their religion sanctioned human sacrifice and gave justification for warfare. Like the Aztecs, they studied the sky for portents, believing that if they did not make appropriate blood sacrifices, the gods would be angry. Even though we don't know very much about who they were, the evidence for these beliefs is clear from what remains of their art. However, for me, the question was not whether the Teotihuacanos were superstitious – they clearly were – but if they were in possession of knowledge that could be of use to us today.

One of the things which has interested me for a long time now is the idea that ancient monuments – and sometimes even whole cities – were oriented in such a way as to reflect the heavens above. This idea was the basis of *The Orion Mystery*, the bestseller I wrote with Robert Bauval. In that book, we put forward the then novel idea that the three great Pyramids of Giza were intended to represent the three bright stars that make up the Belt of Orion, one of the most easily recognisable star groups in the sky. That this is true, I have absolutely no doubt at all, for not only is there a remarkable correlation between these pyramids and the belt-stars but we know from various texts that Orion featured prominently in the Osiris religion of Egypt as the location of their heaven. Hieroglyphic representations of Orion have been found in several tombs of the pharaohs as well as the texts of prayers entreating that the pharaoh's soul be taken to the region of Orion and there united with Osiris, the Egyptian god of the dead, whose star-form was said to be the constellation of Orion.

In mid-1993, while we were still busy writing *The Orion Mystery*, Robert Bauval and I visited fellow author Graham Hancock, one-time East African correspondent for *The Economist*. Hancock's most famous book, *Fingerprints of the Gods*, was not yet finished but he was already famous as the author of *The Sign and the Seal*.[1] Knowing of his interest in Africa and the pyramids, we were keen to discuss with him the possibility of a three-party collaboration on a sequel to *The Orion Mystery*. He latched onto the significance of the Giza–Orion correlation straight away, seeing its wider significance to esoteric research in other parts of the world. 'Have you ever been to Teotihuacan?' he asked us. 'The overall plan is exactly the same. It's clearly based on an Orion correlation. The only difference is that instead of the pyramids being placed along a diagonal, as they are at Giza, the line representing the major axis of the Belt runs through the centre of the Pyramid of Quetzalcoatl and the Pyramid of the Sun, parallel to their western and eastern faces. Look, I'll show you what I mean.' Reaching over to his bookshelf, he pulled down a copy of Peter Tompkins's *Mysteries of the Mexican Pyramids*. There, on pages 238–9, we could see it for ourselves. Ignoring superficial differences of style and the presence of other, lesser buildings, the Orion correlation seemed very obvious. The main difference was, as he said, the orientation of the pyramids.

What was also remarkable was a similarity in comparative sizes. At Giza, there are two giant pyramids (Cheops and Chephren) and one smaller one (Mycerinos). According to Bauval's pattern of correlation, the two giant pyramids represented the two brightest and more southerly stars in the belt: Alnitak and Alnilam. The Mycerinos pyramid, much smaller and offset from the diagonal linking the other two, represented the star Mintaka. Just as this pyramid is smaller and offset, so Mintaka is less bright than

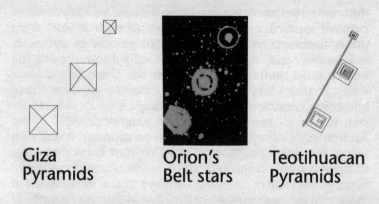

Giza
Pyramids

Orion's
Belt stars

Teotihuacan
Pyramids

Figure 3: Orion correlation at Giza and Teotihuacan

the other two belt-stars and is similarly offset. At Teotihuacan, the situation was less clear-cut. While the Pyramid of the Sun was appropriately large in size and seemed, like the Pyramid of Chephren, to represent the central star Alnilam, the Pyramid of Quetzalcoatl, which according to the theory ought to correspond to the star Alnitak, was at first sight too small to be compared with the Pyramid of Cheops. However, Hancock had an explanation for this too. He pointed out that the Pyramid of Quetzalcoatl sits inside a much larger square enclosure which, for want of a better term, the early Spanish named the 'Citadel'. This enclosed space has exactly the same base area as the Pyramid of the Sun, indicating that esoterically the entire Citadel and not just the little Pyramid of Quetzalcoatl at its centre could be viewed as representing the star Alnitak. In this schema, the Pyramid of the Moon, much smaller in base than either the Citadel or the Pyramid of the Sun and offset from the line linking their centres, could be regarded as the Mexican equivalent of the Pyramid of Mycerinos and therefore of the star Mintaka.

I was at first sceptical of this plan. It seemed too much of a coincidence that, separated as they are by not only distance in space but in time, there could be any link between these two building programmes. Nevertheless, I could not dismiss a cultural link between Egypt and Mexico as entirely fanciful. There was something else about the site that indicated that it, like Giza, was to be viewed as a celestial landscape. The coincidence (if such it was) went much further than just a superficial pattern of pyramids that may or may not have been intended to represent the belt-stars. In the sky, running alongside Orion like a great river, is the Milky Way. Today, because we now have telescopes capable of seeing into deepest space, we realise that it is a galaxy. Like its closest neighbour 'M31' – the Andromeda Galaxy – it is shaped like a spinning wheel with 'arms' comprising millions of stars. We now know that some of these stars are huge, red giants, while others, made up of super-dense matter, are white dwarves. Some, being made of 'dark matter', are completely invisible in the wavelengths of ordinary light. We only know of their presence because they emit electromagnetic waves with frequencies either above or below those of visible light. Many stars are like our own Sun: medium-sized and probably with families of planets orbiting them. Other stars, like Sirius and Alnitak, exist as binary systems. They consist of pairs of stars rotating around a common centre of gravity.

All this detailed information, gained through four centuries of intensive observations using radio as well as optical telescopes, was not available to the Egyptians. They did not realise that our Sun is just one of the many stars that make up our galaxy. In fact, they do not seem to have known that the Milky Way was a galaxy at all. Rather, they conceived of it as being like a river: a celestial counterpart of their own River Nile. On the right bank of this river, not far from the bright star Sirius (which they associated with the goddess Isis), were the stars of Orion. Here, they believed, was heaven. They visualised this heaven as being very much like Egypt, though purer than any earthly place. Ruled over by Osiris, the anthropomorphic god who had once ruled Egypt, they believed this to be paradise. Consequently, they hoped and wished that after death, as resurrected souls, they would ascend to this heaven, where they would live for all eternity with the gods.

The major thesis of *The Orion Mystery*, which Robert Bauval had already explained to Graham Hancock during a previous meeting, was that the Giza complex of pyramids was intended to assist in the ascension of the souls of the

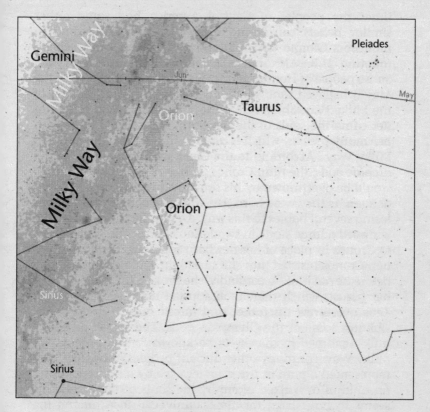

Figure 4: The Egyptian heavenly region

pharaohs – and perhaps the souls of lesser mortals too. By building pyramids representing Orion's Belt on the west bank of the Nile, they were creating an image of their heaven. Then, by enacting important rituals there – such as the 'Opening of the Mouth' ceremony – at astronomically significant times, they believed they could open a portal into heaven. Purified by the rituals and energised by the prayers of the priests, the pharaoh's higher soul would journey from Earth to the stars. Called by the name 'akh' and symbolised as a crested ibis, this soul would become one with Osiris in the same way that Christians hope that after death their souls will become one with Jesus Christ.

Rather surprisingly, one might think, in pre-Columbian

times the peoples of Central America had some rather similar beliefs concerning the Milky Way. According to professors Giorgio de Santillana and Hertha von Dechend (quoting H.B. Alexander's *Latin-American Mythology*), the Sumo people of Honduras and Nicaragua (neighbours of the Maya) believed that 'Mother Scorpion' dwelt at the end of the Milky Way. She received the souls of the dead and from her came the souls of the newborn. She suckled these with her many breasts, which is an idea that echoes the name of Milky Way.[2] According to the Greek myths, the ruler of the archaic gods, the titan Cronos, in fear of a prophecy that he would be overthrown by his son, devoured his own children as soon as they were born. His wife, the goddess Rhea, after losing several babies in this way, eventually tricked him. She wrapped a large stone in swaddling clothes and handed this to Cronos in place of her newborn son Zeus. Perhaps to test her, Cronos insisted that she suckle the child one more time before he ate him. Accordingly, she pressed the stone against her breast, which thereupon spurted milk. As prophesied, Zeus overthrew his father and forced him to disgorge his siblings, some of the Olympian gods. The splash of Rhea's milk remained in the sky to be known forevermore as the Milky Way. For many other ancient peoples, the Milky Way represented a path or track, either used by a hunter chasing an animal or walked along by the souls of the dead. This seems to have been an almost universal idea among the civilisations of antiquity. Even in Egypt, the Milky Way was sometimes referred to as 'the beaten path of stars'.[3]

The Aztecs, perhaps remembering some much older legends, referred to the grand avenue running up the centre of Teotihuacan as 'the way of the dead' and seem to have looked upon it as a terrestrial counterpart of the Milky Way, which was referred to by the same name. This being so, it is thought by some commentators that the individual temples and pyramids along its route were intended, in an abstract way, to represent stars. Unfortunately, although the pyramids, along with the Citadel, can be seen to mirror the stars of Orion's Belt, they are not placed in the right position relative to the 'Way of the Dead' (Milky Way) for the overall correlation to be exactly the same as in Giza. Thus, while this correlation may have been part of the reason for their being built, it is by no means clear that it was the main one or even the most important. Other factors than this are clearly involved.

These sorts of ideas were floating through my head as the next day, following breakfast, we made our way over to the centre of the old city. Once more, I found myself walking up the roughly 4-km Avenue of the Dead between rows of small flanking temples. These were 'talud tablero' buildings: flat-topped structures each of four storeys of progressively smaller platforms, with staircases running up the centre of each side. I could imagine how, during some great fiesta, people would have clambered all over these platforms in order to get a better view of the procession of priests, kings or other dignitaries as they walked along the avenue on their way to the much larger ceremonial buildings at its end.

Walking up the avenue, I was almost immediately set upon by hawkers, anxious to offload all manner of souvenirs from headscarves to whistles to models of the pyramids. Among the goods on sale were figures representing Aztec gods as depicted in the codices.[4] Quite beautifully carved out of the local black obsidian (volcanic glass), intermixed with materials of other colours, they attracted the attention of at least one member of our party. I was about to tell her that they bore almost no relation to the belief systems of the people who built Teotihuacan when I thought again. True, they were the invention of modern craftsmen, but those men were themselves the descendants of the original builders. To say that they were unrelated would be to deny the connection, strong and direct, that exists between the past and the present. For those with the eyes to see it and the hearts to feel it, something of the culture and memory of pre-Columbian America remains accessible even today. These sculptures were a manifestation of that memory. It is in the atmosphere and can be savoured. The tourist knick-knack sellers were, in their own way, part of a bigger picture. For that, they were to be revered. Thus it was that later, on my way back to our hotel, I stopped again and bought a large obsidian screeing-ball from the same man. It sits before me now, black, lustrous and evocative of a culture long gone. But it is not all black: at its two, what one might call polar, ends, there are swirling patterns of white that look like spiral galaxies. The patterns change according to the direction from which they are viewed and the direction of the light source. I take it as a metaphor of Teotihuacan itself, an ancient city with metaphysical connections to the Milky Way.

Halfway up the avenue, I reached a junction leading to the largest building in the whole complex: the Pyramid of the Sun. With a base area similar in size to the Great Pyramid of Giza, it dwarfs everything else in sight. At roughly 228 ft in height, this is not the highest pyramid in Mexico. That honour belongs to the pyramid at Cholula, which, unlike the pyramids of Teotihuacan, is crowned rather incongruously by a Catholic church. Nevertheless, the pyramid of the Sun is an extraordinarily impressive building and was clearly the most important structure in the entire complex at Teotihuacan. I walked over to it slowly, trying to picture how the scene must have looked when Teotihuacan was at its height and its people were celebrating some great festival. This was not as difficult as it might have been, for, this being a Sunday, the site was packed and there were literally hundreds of Mexicans of all ages climbing it. Their brightly coloured clothes and excited voices gave it a carnival atmosphere, in happy contrast to the memory of blood sacrifices that was psychically recorded by the stones. At times gasping for breath in the thin oxygen-deprived air, I struggled my way up the three hundred and sixty-five steps – one for each day of the year – that led to its top. There, to the amusement of some local children, who, having grown up at these high altitudes, were well acclimatised to the thin air, I stood panting like an old dog. However, the effort was worth it for the views. Unlike the many sacrificial victims who must have made the same journey in times past, I was confident I would be able to descend again with my pounding heart still in my chest. In the interim, I was free to enjoy the views and to contemplate what this pyramid might have meant to its builders.

The Pyramid of the Sun is located over a cave system which was probably considered sacred long before the building of the pyramid. This cave system, discovered only in 1971, may in part be of natural formation. The original cave seems to have been formed a million years ago by lava flows. Under the right circumstances, a flow of lava can solidify on the outside to form a tube through which molten rock can flow. If volcanic activity ceases before all of the material has solidified, then the natural flow of the lava can cause the tube to empty, leaving behind an empty space. This must have happened a very long time ago as volcanism in the area has long since ceased. The cave, however, may have proven useful to man from time immemorial. Certainly, it was used for some sort of ritual purposes during the period that the city flourished above it. This is confirmed by votive offerings which have been found inside the cave. If the volcanism which gave rise to the lava flow ever created a volcanic peak, this has long since eroded away. However, one theory for the symbolic meaning of Mexican pyramids is that at least some of them were meant to represent volcanoes. If so, then it seems possible that the Pyramid of the Sun was just such a symbolic replica, made all the more meaningful because it was placed over an extinct lava vent.

Descending from the pyramid, I stood in front of the entrance to the cave, which these days is sealed by a locked wrought-iron grille. Although I tried to persuade a guard to let me in, this proved impossible. Unlike in Egypt, where a fairly modest amount of '*baksheesh*' will normally gain entry to the holiest of holies, he was clearly under strictest orders not to let anyone inside. I had no choice but to accept his argument that this was for my own safety. However, given the way that tourists were allowed to climb all over the pyramid, this argument seemed weak. It seemed more likely that, as scientific research on this cave system was and still is ongoing, the authorities did not want visitors contaminating whatever evidence may still exist of its early use and purpose. Be that as it may, there was plenty of evidence above ground for Teotihuacan's volcanic past. Unlike the Great Pyramid of Giza, the Pyramid of the Sun is not made out of solid stone blocks. It is in fact an earth mound, shaped into five major steps and faced with small pieces of petrified lava. This building material, in varying shades of red, brown and black, is used throughout Teotihuacan, all its major buildings being similarly constructed.

A major resource of the Teotihuacanos was obsidian. While today this is primarily used for the making of tourist goods, such as my crystal ball, in ancient times it had other uses. When highly polished, sheets of obsidian were used for the making of mirrors. One such mirror found its way into the hands of the celebrated Elizabethan magus Dr John Dee. He used it for 'screeing': that is, looking into the future. Today, it is in London in the British Museum and can be viewed there, along with other occult artefacts of his such as his crystal ball (used for invoking angels) and various magical seals. His mirror is highly polished and deep black in colour. It gives a whole new meaning to the phrase 'looking into a glass darkly'. The Aztecs (and probably the Teotihuacanos before them) had similar obsidian mirrors and may have used them in a similar way to Dr Dee. Certainly, their naming of one of their most prominent deities *'Tezcatlipoca'*, or Smoking Mirror, reflects the importance of such mirrors to their culture. In their mythology, he was both the brother and adversary of the better-known hero-god Quetzalcoatl. Whereas the latter was associated with water, fertility and wind, his dark brother symbolised conflict. Images show him with an obsidian mirror attached to the back of his head and sometimes with a second mirror in place of one of his feet. He was both feared and worshipped by the Aztecs, who seem to have believed that they were his slaves.

Like ordinary glass, obsidian is extremely sharp when fractured but with the added bonus that it is much harder. Thus, as well as for mirrors, the Teotihuacanos used it for making the blades of sacrificial knives. For this purpose, it was ideal; though fragile, obsidian blades are capable of cutting through flesh more efficiently than steel.[5] These were not its only uses. In the hands of a master-craftsman, obsidian can be chipped to make all manner of bladed tools, including the heads of arrows and spears. So important was obsidian to the peoples of Mesoamerica, who were still living in the Stone Age at the time of the Spanish conquest, that the Teotihuacanos traded it throughout the region. Obsidian knives of the characteristic Teotihuacan type have been found hundreds of miles away from the city in the Maya heartlands. These, and the presence of what are now recognised as Mayan neighbourhoods in Teotihuacan itself, indicate the central role the city played in the life of the Maya throughout the Classic period. It would seem that it was viewed by them not just as an emporium for buying useful tools but rather as a centre of religious pilgrimage.

Many shards of broken obsidian mirrors have been found inside the concealed cave under the pyramid, perhaps indicating a connection to a secret cult. Whatever is the truth of that, the fact that the Teotihuacanos built their largest pyramid right over the cave complex indicates that the underground location was of great symbolic importance. Comparisons would be with the way the Church of the Nativity at Bethlehem is built over the cave believed to be where the baby Jesus was born or how the Church of the Holy Sepulchre in Jerusalem is reportedly built over Jesus's tomb. In a similar vein, many archaeologists and historians are now convinced that the cave complex under the Pyramid of the Sun is intimately connected to the myth of Quetzalcoatl, one version of which says he sacrificed himself at that very spot so that the Sun of our present age would move through the sky.

Leaving the Pyramid of the Sun, I resumed my march along the Avenue of the Dead. This terminated with the Plaza of the Moon: a city square flanked by rather larger buildings than those along the approach road. Immediately ahead was the Pyramid of the Moon, similar in shape to the Pyramid of the Sun but much smaller in volume. Again, I climbed to the top of the pyramid, and there took out from my pocket the Global Positioning Systems (GPS) satellite navigator that I had brought with me. It was not that I was lost or unable to find my bearings. Rather, I was keen to check for myself the precise orientation of the great processional way, the Avenue of the Dead, which seemed to reach its destination with this pyramid. GPS works by establishing connection with purpose-built satellites as they circle in the sky. There are 24 of these satellites, launched and maintained by the United States. Though their prime function is military, they also send out tracking signals that can be received by anyone with a GPS device. As long as the device can pick up signals from at least four satellites, it can compute a relatively accurate bearing for itself, giving the position north or south of the Equator and east or west of the Greenwich meridian. It took a few minutes for the GPS to settle down and link to all the satellites in the neighbourhood.

Unfortunately, so that enemies cannot use it as they themselves do for missile-guidance purposes, the American military has made the civilian GPS system less accurate than it might be. Consequently, the stated position tends to vary over time, oscillating from one extreme to another about a central point. To compensate for this, I took several readings and then averaged them. The final reading, possibly not exact to the last decimal place, came to 19° 41.969' north and 98° 50.634' west. Later, I made a second set of measurements a few kilometres down the Avenue, where it came to its end close by the hotel. The readings I got here averaged out to 19° 40.769' north and 98° 50.996' west. By subtracting one reading from the other, I could see that the Pyramid of the Moon was 1.20 minutes north of the hotel position and 0.362 minutes east. This means that facing towards the Pyramid of the Moon, within the limits of accuracy supplied by the method of surveying the site using GPS, the Avenue of the Dead is oriented at an angle of roughly 16° 47' east of north. This was slightly different from the figure of 15° 25' east of north that is frequently quoted in books and on websites, but not all that much.

I had hoped that the discrepancy would prove to be greater than this, as it would have explained the generally accepted association of the Way of the Dead with the Milky Way. As is evident to any observer when the sky is clear and dark enough, our galaxy forms a wide band of stars. Because the Earth is rotating and orbiting the Sun, the elevation and orientation of this band varies depending on the time of night and day of the year. Prior to coming to Mexico, I used a computer program called Skyglobe to ascertain when, at the latitude of Teotihuacan, the Milky Way would reach its greatest elevation. This is when it gives the appearance of a white ark that spans the sky from horizon to horizon and passes directly overhead through the zenith point.[6] Because the axis of the Milky Way is not coincident with the north–south axis of the Earth, the Milky Way appears to rise and fall in the sky as the celestial sphere appears to turn (making one complete revolution in 24

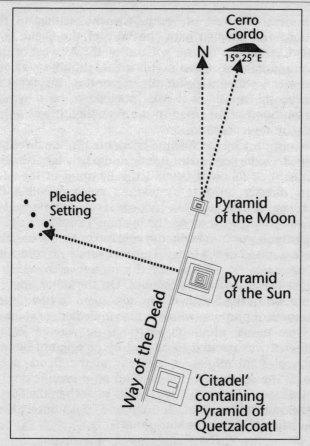

Figure 5: Overview of Teotihuacan

hours), with first the northern 'limb' of it raised and then the southern. As seen from Teotihuacan, when one limb is raised it forms an arc of stars running from north-east to south-west and the constellation of *Aquila* is near the zenith. When the other limb is raised, the Milky Way stars run from north-west to south-east and the constellation of Orion is near the zenith. I discovered that during the period of the city's greatness (*c.* 100 BC–*c.* AD 750) the northern limb would be raised, with the median plane of the galaxy crossing the horizon at roughly 45° west of north and 45° east of south. Conversely, the southern limb crossed at roughly 45° east of north and 45° west of south. Clearly, neither of these positions was aligned with the Way of the Dead, which points to positions on the horizon of 15° 25' east of north and the same west of south. This meant that the Way of the Dead was not meant to literally mirror the Milky Way as it transited the zenith. If it was intended to be a symbolic representation of the 'road in the sky', then it was only an approximate representation.

The reason why the Avenue is oriented in the direction it is appears to be partly astronomic and partly terrestrial. The alignment, if followed beyond the Pyramid of the Moon, points directly towards an extinct volcano called *Cerro Gordo*, or Fat Mountain. Indeed, as I approached the pyramid and looked along the avenue, I could see the way the pyramid was framed by the volcano behind it as though it were a model of the latter. This was surely no coincidence and seems to be evidence of the reverence in which the ancient Mexicans held volcanoes. On the other hand, the orientation of the Sun Pyramid does seem to have had an astronomical purpose (which is also indicated, of course, by its Aztec name, although we have no way of knowing whether it was referred to as that by its original builders). Facing at right angles to the Way of the Dead, it looks towards the setting point of the Sun on a specific day. This day (in *c.* AD 300, when it was built) was when the Sun was in the constellation of Taurus and positioned directly underneath the Pleiades star-group.

M45, or the Pleiades star-cluster, is probably the most easily recognised of all asterisms. An open cluster of hundreds of stars, there are seven that can be distinguished with the naked eye. These were known to the Greeks and Romans as the 'Seven Sisters' and were called the *cabrillas*, or goat-calves, by the Spanish conquistadores. We know from post-colonial sources, such as the works of the friar Bernadino Sahagun, that the Pleiades played an important part in Aztec rituals at the end of every 52-year 'jubilee'. The significance of the period of 52 years will be explained in the next chapter, but these rituals always took place in November on the day when the Pleiades crossed the southern meridian at midnight:

The measurement of all times that the Indians [the Mexicans] carried out was as follows: the largest measurement of time was of 104 years, and they called it a century; half of this period, which is 52 years, they called a sheaf of years (*gavilla*). This number of years they have counted since ancient times; it is not known when it began, but they quite firmly believed that the world would come to an end at the termination of one of these sheaves, and they had prophecies and oracles that the movements of heaven would cease then, and they took for a signal the movement of the *Cabrillas* [Pleiades] in relation to the night of this feast, which they called *Toxiuh molpilli*; for on that night the *Cabrillas* were in the middle of the sky at midnight, corresponding to this Mexican latitude.

On this night, they lit the new fire . . . and all the satraps and ministers of the temples went in great procession and solemnity . . . from here, from the Temple of Mexico [Templo Mayor] early at night; they went to the summit of that hill near Ixtapalapa and which they call Uixachtecatl . . . to where there was a solemn *cu* [small temple] made for that ceremony . . . and when they saw that they [the *Cabrillas* or Pleiades] passed the zenith, they understood that the movement of the heavens was not to cease, and that it was not the end of the world, but that it would last another 52 years, and that the world would surely not come to an end. At this hour, great multitudes were in the hills surrounding this province of Texcoco, Xochimilco and Quahtitlan, waiting to see the new fire, that was the signal that the world would continue on; and as the satraps brought out the fire with great ceremony in that hill's *cu* it then appeared all around the hills, and when those who were there saw it, they were so happy that they yelled and their yells reached up to heaven, for the world was not coming to an end and they undoubtedly had 52 more years.[7]

I wrote about these fire festivals at some length in *The Mayan Prophecies*, pointing out that they were not unique to the

Aztecs but were celebrated by the Maya as well. At around the same time as Bernadino Sahagun was recording all he could concerning the ancient beliefs of the Aztecs, a group of elders of the Quiché Maya of Guatemala were recording their own traditions for posterity in the *Popol Vuh*. Written in Quiché using the Spanish alphabet instead of their own hieroglyphic signs, it appears to be largely a transcription of a long-vanished bark-book. As we will see later, it too contains references indicating how important the Pleiades were considered to be.

We do know (because Sahagun records it) that the last Aztec fire ceremony to be held prior to the arrival of the Spanish and the suppression of such pagan customs was in 1507. At that time, and indeed ever since, the Pleiades as seen from the region of Mexico City (which includes nearby Teotihuacan) were too low in the sky to culminate at the zenith. However, between roughly AD 700 and AD 750, anyone standing in Teotihuacan would indeed have seen the Pleiades culminating exactly overhead at midnight on 19 or 20 November. If we count back in periods of 52 years from 1507, there should have been a festival on or around AD 727. From this time until 2 cycles later (i.e. the completion of an Aztec century of 104 years in AD 831), the Pleiades continued to culminate overhead at Teotihuacan. It is interesting, therefore, that it is during this period that archaeologists believe the city was finally abandoned.

Standing on the top of the Moon Pyramid, I thought about this. Looking backwards in the direction of Cerro Gordo, I remembered the story of the Aztecs and their festival of the New Fire. It was not hard to imagine similar festivals taking place here centuries earlier. On this holy day, all fires throughout the city would be extinguished and the people would prepare themselves for the great event. Then, during the afternoon and early evening, some priests, probably accompanied by sacrificial victims, would make their way to the top of the sacred volcano of Cerro Gordo. The rest of the people, thronging the avenue and every available platform, would stand either in hushed silence or saying their prayers. All eyes would be turned towards the mountain top, the centre and focus of their attention. Then, just as the Pleiades reached their highest position in the sky, the priests on the crest of Cerro Gordo would light a beacon-fire. The sacrificial victims, who may even have been volunteers, would either jump or be thrown into the flames, just as the gods had apparently done at the start of the present age. As the flames took hold, so the beacon would become visible to the people standing waiting in Teotihuacan itself. A great cheer would go up from the crowds as, looking to the heavens, they saw the Pleiades moving from the zenith. Brands of this 'new fire' would be plucked from the blaze and taken back to the city by runners. Another great cheer would go up from the crowds as the fire arrived. It would be used to light other fires, perhaps contained in braziers on top of the Pyramids of the Sun and Moon.

It would have taken several hours for the runners to get back to the city and by now the heavens would have changed. Looking at the sky, all would see that the Pleiades had moved onwards from their high position at the zenith and were now setting in the western sky. This would be most clearly visible to the priests who were standing on top of the Pyramid of the Sun. Towards the east, the Sun would be rising, its first rays beginning to obliterate the stars. To ensure that it would rise, the priests would probably sacrifice other victims on top of the pyramids, throwing their bodies into the fires. If the Teotihuacanos were anything like the Aztecs (and it is becoming more and more apparent that they were), then as the Sun continued to rise there would have been great rejoicing. Torches taken from the funeral pyres on top of the pyramids would have been distributed among the people and they would have used them to light their domestic fires.

Leaving the summit of the Pyramid of the Moon, I made my way back down to the plaza at its foot. Near to here was a building labelled as a palace, though it seemed to me more likely that it was a religious building – perhaps a place of initiation – rather than a house. The building itself featured an open courtyard with a 'cloister' surrounding it. The roof of this was supported by pillars, ornately carved with representations of a curious hybrid of bird and insect referred to by archaeologists as the Quetzal Butterfly. What this curious symbol signified was not explained in the local guidebook but it clearly relates to the idea of metamorphic rebirth; that we live this life in a world of caterpillars and after death have the possibility of resurrection into a 'higher world' of brightly coloured butterflies and birds.

As a symbol of resurrection, the butterfly (or alternatively dragonfly) is not confined to Central America. It is met with, for example, on the famous Ring of Nestor that was found in eastern Greece in 1925. However, butterflies, especially monarch butterflies, have a special relevance in Mexico. Unlike their European counterparts, which appear to live totally isolated lives, these butterflies make an annual migration from all over North America to the forested hills of Mexico. Because their bodies contain toxins harvested from plants, they are left largely unmolested by birds. Thus, they gather in their millions in just a few valleys. This must have been known to the builders of Teotihuacan and no doubt the butterfly valleys were considered sacred too. The heirs of these butterfly-worshippers (if that is what they were) are the people of the great Valley of Mexico, where today stands Mexico City: one of the most densely populated cities on Earth. Returning here would be my next destination.

CHAPTER 3

The Rolls of Time

On the second day of my return to Mexico City, I went to see the Aztec calendar stone, made famous by Humboldt and Lubbock and now kept in the Museo Nacional de Antropología de la Ciudad de Mexico. Since 1964, this museum, the largest and most authoritative in Mexico, has been housed in Chapultepec Park. The building, designed by Pedro Ramirez Vasquez, is impressively modernistic. With exhibition rooms arranged around an open quadrangle, it reminded me of a cathedral cloister. The resemblance, though, was functional rather than architectural. There is nothing Gothic, still less baroque, about this structure, which is minimalistic in the extreme.

In a curious way, the sheer modernity of the building seems entirely appropriate for the semi-abstract artworks it contains. The central area of the courtyard of the cloister was shielded from the elements by a gigantic 'umbrella' (the architect's own description), supported by a single column at its centre. Perhaps as a play of opposites, the shaft of this umbrella acts as a fountain, with water cascading from it into a pond below. With so much water flowing through the air, the effect is to cleanse the atmosphere. The sheer modernity of the building has another effect, which is to anchor the visitor in the modern world whilst viewing the ancient art of pre-Columbian Mexico. This, I found, was just as well, as much of it, featuring as it does images of human torture and sacrifice, is gruesome in the extreme. By creating a modern environment for such objects as the obsidian knives used for sacrificing victims, a welcome distance was created between object and viewer.

Just inside the entrance to the Aztec rooms was a beautifully carved, curled-up rattlesnake. Much larger than life-sized, it was a reminder of just how important the cult of the serpent was to not only the Aztecs but all their neighbours in Central America. Nearby was a large square block of stone sculpted to form a bowl. Like other, rounder bowls in the room, it was no doubt used as a receptacle for the still-beating hearts of sacrificial victims. I tried to put such thoughts aside as I examined its intricately carved sides. One of these showed a mythological bird sitting on top of a tree and eating its fruits. It reminded me then of the Aztecs' myth of foundation, today the subject of the flag of Mexico: an eagle standing on a cactus while clutching a serpent in its beak.

The Mexica, as the Aztecs called themselves, were fierce warriors. Within a few years of their arrival, they established themselves as the major power in the area. They made alliances with the neighbouring cities of Texcoco and Tlacopan and together they set about building what was to become the Aztec Empire. They also absorbed local customs and traditions with regard to art, pyramid building and religious concepts: above all, the need for human sacrifice if the gods were to be placated. Some of these gods, such as Tezcatlipoca, 'Smoking Mirror', were indigenous to the area. Others, such as *Huitzilopochtli*, or Left-hand Hummingbird, and his mother *Coatlique*, or She of the Serpent Skirt, were purely Mexican.

Coatlique herself was the subject of a truly terrifying cult. Her cult statue, which was hastily buried by the Spanish authorities soon after the conquest, was rediscovered in 1790. By then, the Aztec Empire was sufficiently removed in time for such an object to be treated as an interesting relic of a bygone age and not be destroyed as an obvious threat to the Catholic religion. Today, it stands in the heart of the museum. Partly shrouded in darkness and horrifying as it is to look at, it gives barely a glimmer of the horrors perpetrated in her name.

I remembered the statue from my previous visit and, not having a decent photograph of this, one of the treasures of the museum, went over to examine it in some detail. It reminded me of a Frankenstein-like horror story I had once read. In this story, a sorcerer had assembled a giant heap of body parts, not all of them human in origin, and whispered some magic spells over this assorted carnage. Immediately, a demon entered the heap and brought it to life. Forming its parts into the semblance of a body, it set about terrifying the neighbourhood. Eventually, after pages of unpleasantness, the hero found the means to exorcise the demon and the rotting flesh collapsed back into an inanimate heap. Once more, good had triumphed over evil and I could feel safe turning off the light and getting into bed.

Looking at the statue of Coatlique was like looking at a representation of the animated heap. Her 'head', if one could call it that, was composed of two opposing serpents' heads, their eyes looking forwards and fangs bared. Her body and skirts were composed of writhing serpents and her feet were the claws of a jaguar. Worst of all, she had a necklace of severed human hands, hearts and a skull hanging round her neck. Though often referred to as an Earth-mother goddess, it was clear that she had much more in common with the Indian goddess Kali than the comforting Demeter of Greek mythology. The sheer horror of this statue, which must once have been smeared with human blood and would therefore have reeked of putrefaction, was brought home to me when I later read the account given to Friar Bernadino Sahagun of Aztec festivals. The thought that this piece of stone had born silent witness to atrocities such as heart-sacrifice and live flaying filled me both with sorrow and curiosity. What was it, I wondered, that had convinced the Aztecs that their gods required them to behave in such a cruel way? If it was fear, then what was it that they were afraid of? Surely not this idol itself?

Of course, the dumb statue of Coatlique could not give a reply to this question, but there was something else nearby which did provide some clues: the famous Aztec calendar stone. I had seen it before but had forgotten just how large and impressive it really is. Carved in 1479, when the Aztec Empire was at its height, it is a very detailed sculpture that at first confuses the onlooker with its complexity. However, careful examination of its symbolism reveals that at core it illustrates the same teachings concerning time and ages that are to be found in certain books written by the Aztecs after the conquest. At the very centre of the stone wheel is a human-looking face, thought to represent the Aztec sun god, Tonatiuh. At either side of this face are eagle's claws. They are shown clutching the human hearts that, according to Aztec mythology, he had to be fed in order for the Sun to keep moving through the sky. This is also why Tonatiuh has his tongue protruding: it probably indicates that he is thirsty and must therefore be given human blood.

But the calendar stone is much more than a reminder of the bloodlust of the Aztec gods. It is important because it embodies their belief system concerning previous ages and the ongoing cycle of time. Since they held many of these ideas in common with the Maya, it is worth analysing in some detail for the clues it holds to the understanding of the Mayan calendar and their prophecies for the ending of the current age.

THE AZTEC CALENDAR

The Aztecs, like all the other nations of Central America, made use of a 260-day cycle which they called the *Tonalamatl*. This cycle was not calendrical, in the normal sense, as it was not directly related to the movement of the Sun or Moon.[1] The cycle of 260 days was generated artificially by counting 20 day-names against the numbers 1 to 13. Looking closely at the Aztec calendar stone and comparing it with a guide that I bought in the museum shop, I could see that the day-names were depicted by logographs. They were shown on the calendar stone making up a continuous ring or cycle around the 4 *Ollin* symbol. They are:

Cipactli	Crocodile
Ehecatl	Wind
Calli	House
Cuetzpallin	. Lizard
Coatl	Serpent
Miquiztli	Death
Mazatl	Deer
Tochtli	Rabbit
Atl	Water
Itzcuintli	Dog
Ozomatli	Monkey
Malinalli	Coarse grass
Acatl	Reed
Ocelotl	Jaguar
Cuauhtli	Eagle
Cozcacuauhtli	Vulture
Ollin	Movement
Tecpatl	Obsidian Knife
Quiahuitl	Rain
Xochitl	Flower

These day-names were counted in repetition against the numbers 1 to 13 in rather the same way that we count our days of the week – Sunday, Monday, Tuesday, etc. – against the numbered days of our months. After the number 13 was reached, they would go back to 1 and pair it with the next symbol on the cycle of 20 day-names. Another way of thinking of this is to regard the two cycles as being like interlocking gear wheels.

Thus, if day one was '1 Crocodile', then day two was '2 Wind', day three was '3 House' and so on up to '13 Reed'. The number cycle having reached its end, the next day would be '1 Jaguar'. This would be followed by '2 Eagle' and so on up to '7 Flower'. At this point, the day-names would run out and have to restart. Thus, the day following '7 Flower' would be '8 Crocodile'. This would be followed by '9 Wind'. When we get to the 260th day, in this case '13 Flower', the cycle repeats itself by beginning again with '1 Crocodile'. By counting in this way, it becomes evident that the same combination of number and name will not recur for 260 days (13 x 20). Thus is produced the Tonalamatl, or sacred calendar of the Aztecs.

Day-names
ring

Figure 6: Interlocking Aztec wheels

In addition to their curious 260-day calendar, the Aztecs had another calendar, a 'Vague Year' of 365 days, which they called the *Xiuhpohualli*. This consisted of eighteen 'months' of twenty days' duration followed by a short month of only five days. In this case, the days of each month were simply counted consecutively, the first being the 'seating' or day 0 of the month. This was followed by the numbers 1 through 19 (or 0 through 4 if the month had only five days). Thus, the days were given in order, e.g. 0 *Quecholli*, 1 *Quecholli*, 2 *Quecholli*, etc.

Since my last visit to Mexico, I had read more of the works of Bernadino Sahagun, the Franciscan friar who preserved much concerning the traditional knowledge of the Aztecs. He writes that each of these 20-day months was dedicated to a particular god (or gods) and was associated with at least one feast. These usually involved human sacrifice. According to him, the names and dedications were as follows:

Month	Start date	Name	Deities
1	2 February	*Atlcahualco* 'Departure of the Waters'	Feasts of the god *Tlaloc* and goddess *Chalchiuhtlique*
2	22 February	*Tlacaxipehualiztli* 'Flaying of Men'	Feast of the god *Xippe-Tototec*
3	14 March	*Tocozontli* 'Lesser Vigil'	Feasts of the god *Tlaloc* and goddess *Coatlique*
4	3 April	*Veytococoztli* 'Greater Vigil'	Feasts of the god *Cinteutl* and goddess *Chicomecacoatl*
5	23 April	*Toxcatl* 'Dryness'	Feasts of the gods *Tezcatlipoca* and *Vitzilopochtli*
6	13 May	*Etzalqualiztli* 'Meal of Corn and Beans'	Feast of the water gods called *Tlaloques*
7	2 June	*Tecuilhuitontli* 'Lesser Little of the Lords'	Feast of the goddess *Vixtocioatl*
8	22 June	*Veytecuilhuitl* 'Greater Feast of the Lords'	Feast of the goddess *Xilonen*
9	12 July	*Tlaxochimaco* 'Birth of Flowers'	Feast of the god *Vitzilopochtli*
10	1 August	*Xocotlvetzi* 'Fall of Fruit'	Feast of the god *Xiuhtecuhtli*
11	21 August	*Ochpaniztli* 'Sweeping the Roads'	Feast of the goddess *Teteuinna*
12	10 September	*Teotleco* 'Return of the Gods'	Feast of the god *Tezcatlipoca*
13	30 September	*Tepeilhuitl* 'Feast of the Hills'	Feast of the god *Tlaloc*
14	20 October	*Quecholli* 'Precious Feathers'	Feast of the god *Mixcoatl* and others
15	9 November	*Panquetzaliztli* 'Raising of the Banners'	Feast of the god *Vitzilopochtli*
16	29 November	*Atemoztli* 'Fall of Water'	Feasts of the water gods called *Tlaloques*
17	19 December	*Tititl* 'Stretching'	Feast of the goddess *Illamatecutli*
18	8 January	*Izcalli* 'Resuscitation'	Feast of the god *Xiuhtecuhtli*
19	28 January	*Nemontemi* 'Empty Days'	No festivals during these five unlucky days

I won't here go into the details of the way the Aztecs marked these feasts. Suffice it to say that these were nearly always accompanied by gruesome rituals involving heart-offerings, beheadings, flayings, live immolation and even cannibalism. Indeed, reading through Sahagun's account, which is delivered factually and generally without comment, is in itself a disturbing experience. The good friar lets the appalling fact of a civilisation in thrall to the most cruel rituals speak for itself.

There is, however, one annual ritual that I find of more than passing interest from a calendrical point of view. This festival, in honour of Xiuhtecuhtli, a fire god with strong solar connotations, was celebrated during the month of Xocotlvetzi. As this ceremony is of interest to what follows later in this book, I will reproduce most of what Sahagun himself has to say about it.

> The tenth month was called Xocotlvetzi. As soon as the festival of Tlaxochimaco [held the previous month] was over, they cut a big tree in the forest, 25 fathoms in height (150 ft); they cut off all the twigs and branches, only leaving the tender shoot at the top. Then they cut down other (smaller) trees, hollowed them to form a sort of litter, placed the felled tree on these, tied it with ropes, and as they dragged it (from the forest) it never touched the soil, nor was its bark scratched because it rested on these bent or hollowed-out trunks. As they approached the town, the ladies and prominent women came to meet them. They carried bowls (*jicaras*) of cocoa, that the bearers of the tree might drink, and also flowers, with which they decorated the men who brought the tree. As soon as they reached the court of the temple, the *Tlayacanques*, or leaders, began to shout with all the power of their lungs to have the people assemble in order to help lift that tree (rather, trunk), which they called *Xocotl*. As soon as all the people had arrived, they tied the ropes around the tree and, after having dug a hole in the place where it was to stand, all tugged at the ropes amid great shouting until the tree stood erect. They then closed the hole with rocks and

earth and thus the tree stood for 20 days. On the eve of the festival called Xocotlvetzi, they felled it again, lowering it little by little to prevent it breaking or cracking. They rested it on some logs tied two and two together, which they call *quahtomacatl*, and thus they laid it down without injury. There they left it and went away. The ropes were left tied to the tree. Thus it remained all that night. At dawn of the feast day, all the carpenters came together, bringing their tools, and tooled the tree until it was very straight and, taking off any curb or knot, they succeeded in getting it very even. They then prepared another pole of five fathoms (30 ft) in length and curbed it, to put it against the top of the tall tree where the shoots of the crown began. These shoots or branches were gathered in the curb of the other pole, where they were tied with a rope, which was wound from there closely around the crown and down again to the juncture.

Now that this was done, the priests, dressed in their vestments and ornaments, adorned the tree with papers and were assisted in this by the so-called *Qüaqüaciviltin* and the *Tetlepantlaz*, who were three very tall men. One of these was called *Coicoa*, the second *Cacancatl*, and the third one *Veicamecatl*. They put these papers on amid great noise and much solicitude. A statue resembling a man and made of dough of wild amaranth-seeds was also adorned with papers. The paper used was pure white without paint or dye. On the head of this statue they fastened cut papers which looked like hair; on both sides they fastened paper-stoles reaching from the right shoulder to the left armpit. On the arms they put paper fashioned like waves, on which were figures of the sparrow-hawk. He also received a belt of paper. Above this other papers placed back and front simulated the *vipil* (blouse or shirt). On the sides of the statue and on the tree, from the feet of the former down to about the middle of the tree, long papers were hung, which waved in the air. These latter were about one half-fathom (3 ft) wide by a length of ten fathoms (60 ft), more or less. They also placed three large tamales made of wild amaranth-seed on the head of the statue, fastened on three sticks. When the tree was adorned with all this, they tied ten ropes about its middle, and then pulled, with big shouts encouraging one another to heave simultaneously. As it was being lifted, they shoved logs, always two and two tied together, under it, as well as props to rest on as it rose slowly. A great shout was given and much shuffling of feet as the tree stood erect, and at once large stones (rocks) were put against it with earth on top, so that it could not fall. Then everybody went home, no one remaining in the court.[2]

I will spare the reader the lurid description of how captives were then sacrificed to Xiuhtecuhtli, the god of fire and time, as it is too awful to contemplate. However, there are a couple of details worth considering. Before the captives were sacrificed, the warriors danced with them. They stopped dancing at sunset (i.e. when the Sun was in the west) and at midnight each warrior cut a lock of hair from the crown of his captive's head. This is significant as the Sun would then be at the lowest point of its cycle, emphasising the connection between the ceremony and the motion of the Sun. The following day, shortly after dawn (when the Sun was in the east), the unfortunate captives would be sacrificed to the god of fire. Again, it seems, timing was all-important.

These sacrificial ceremonies completed, the assembled Aztecs would go home to eat before the next stage of the ritual. For once, this does not seem to have involved the spilling of blood but was in the nature of a competition to see which of the young men was the fittest.

After that (dinner), all the young men, youths and boys would gather; all those who wore long locks of hair in the neck, which they called *cuexpaleque*, and all the rest of the people met in the court of Xiuhtecuhtli, in whose honour this feast was held, and at noon they began to dance and sing, women dancing in an orderly manner among the men. The court became so filled with people that there was no way to get out, and they were all greatly crowded. When they got tired of singing and dancing, they gave a great shout and left the court and went to the place where the tree (pole) had been erected, and the roads were so very full of people that they jostled against one another. The captains of the young men were (stationed) around the tree so that nobody should go up until it was time; they defended the ascent with blows of their clubs, and the youths, determined to climb the tree, shoved those who defended the ascent out of the way with fist-blows and, taking hold of the ropes, began to climb. Along each one of the ropes a great many youths climbed up in competition; on each one of these ropes hung a cluster of young men, determined to get to the top. Yet, though many attempted the climb, only a few succeeded. The first one to reach the top took hold of the idol (statue) made of dough and wild amaranth-seed, which had been fastened there, took from it the shield and the darts and the spears with which it was armed as well as the instruments with which to shoot off the shafts (spears) and which is called *statl*. He also grabbed the tamales on the sides and broke them into pieces over the heads of the people (gathered below). And all these people looked upwards as the pieces fell for them to catch; some quarrelled and cuffed one another in order to get a few morsels. There was great shouting before they obtained all that was thrown to them. Others (of the youths who had climbed to the top) took the tufts which were on the head of the statue, and he who had taken them threw these down too. This done, he who had climbed up first came down with the weapons he had taken; as he reached the bottom they received him with great applause and carried him, and took him to the top of the cu (temple) called *Tlacacouhan*. There many (of the) old men gave him jewels or devices for the valour he had shown. Then they all pulled at the ropes with great force and threw the tree down on the soil. It caused a great clash and broke in pieces. After this, they all went home and no one remained there.[3]

On reading this account, I was curious to note that the dancing in the courtyard of the temple of Xiuhtecuhtli began at noon, when the Sun was culminating in its highest position in the sky. This, I could see, was important and clearly echoed in some way the Festival of the New Fire, which I talked about at some length in *The Mayan Prophecies*. However, there was clearly more to it than this. Reading extensively on the subject of Aztec rituals and their connection with the drama of the ending of the present age, I realised that one key to the understanding of this curious pole ritual is to be found on the first page of the Aztec *Fejérváry-Mayer* codex. In this colourful representation of Aztec beliefs, the god Xiuhtecuhtli, armed with darts and spears, is shown in the centre of what one can only call a mandala. Arranged around him, according to the directions

The *Fejérváry-Mayer* codex

of the compass, are four trees. Sitting on top of each of these trees is a bird. They reminded me of the sparrow-hawk decorations on the idol that Sahagun says crowned the Xocotlvetzi pole. It would seem, therefore, that this idol must have been a representation of Xiuhtecuhtli himself and that the winning of his arms by the first young man to climb the pole was looked upon as a special blessing.

Sahagun's description of young men climbing a tall pole reminded me also of a ritual that I saw enacted on my previous visit to Mexico. Called the *Danza de los Voladores*, it originates with the Totonac Indians of Papantla, a city nearly 200 miles north-east of the former Aztec capital. For this ritual, a group of five brightly dressed young men ascended an 80-ft pole. At the top of the pole was a tiny platform, barely 8 in. square. Attached to this and standing out possibly a yard or so from the trunk of the pole itself was a square rotatable frame. The first of the men to ascend the pole stood on the small central platform and started to play a bamboo flute and drum while dancing on the 8 in. plate. He was soon joined by the others, who took their positions around him, sitting on the movable frame, to which they attached themselves with ropes. Then, in unison, they fell backwards from the frame to float in the air like so many birds flying round a tree. The ropes being of precisely the right length and wound around the top of the pole the correct number of times, they made exactly 13 circulations before coming down to earth and landing on their feet. Then, with all the dignity of an accomplished acrobat, the musician climbed down the post to join his friends in receiving applause from the gathered audience.

It was evident to me at the time that this dance was done in memory of a ritual that must once have had a deeper message than one of daring acrobatics. Reading Sahagun's account, I could see that, though they differed in some respects, this Totonac custom had to be related to the Xocotlvetzi ritual of the Aztecs. More than that, it seems to amplify Sahagun's account by providing details that were missing from the Aztec ritual. It became clear to me that the Totonac ritual must have originally been calendric in nature. The four flyers must represent the four phases of the day, the four seasons of the year and the four directions.

Furthermore, the 4 x 13 circulations of the pole made by the flyers echoes the number of weeks in a year: 52. Could this be chance? I didn't think so, especially not as the number 52 features prominently in the calendrical system of the Aztecs and is a natural product of their system of counting years.

As we have seen, the Aztecs used 2 entirely different systems for counting the days, the first being the 260-day Tonalamatl and the second the 365-day Xiuhpohualli. It follows from this that each day has two names: one generated by the Tonalamatl cycle and the other according to the Xiuhpohualli. Thus, when writing dates, the Aztecs recorded both names: e.g. 4 *Xochitl* 8 *Izcalli*. This has an interesting result, as the same combination of the 2 names will not recur for 18,980 days. This period was known as a gavilla or 'sheaf' of years and equals both 73 x 260 and 52 x 365 days.

The gavilla mattered very much to the Aztecs, for whom it was the equivalent of a biblical jubilee. At the end of each 52-year period, they would hold the new-fire festival described in the last chapter. The fact that they used observation of the Pleiades crossing the southern meridian at midnight to tell them when the new year should start gives a clue as to how they kept their calendar accurate with the stars. As we all know, the true length of the solar year is about a quarter of a day longer than 365 days. To keep our calendar in sequence with the solar year, we add an extra day every fourth or 'leap' year and make a further adjustment of a day at the end of some centuries. The Aztecs didn't do this, and in any case this would have proved difficult to do while maintaining the integrity of the connection between the two calendars. As a consequence, after 52 years the Xiuhpohualli calendar would have been ahead of the real solar year by 13 days. The obvious solution to this awkward lack of synchronicity would, I believe, have been to wait for the end of a gavilla of 52 years and then suspend both calendars. By pausing the ordinary flow of the calendars for a period of 13 days, they could give the sky time to 'catch up'. They could then begin a new sheaf of years with the calendar's integrity re-established and without breaking the link between the Tonalamatl and the Xiuhpohualli cycles.

I don't know if the Aztecs actually did this, and the Aztec calendar stone doesn't appear to give any clues. However, it seems to me the best solution to keeping their 'vague' years of 365 days in synchronicity with the Earth's true year of roughly 365.25 days without damaging the repetitive cycles of the Tonalamatl calendar. Certainly, without some method for correcting the calendar, their 'new year' would have advanced through the year by 1 day every 4 years, taking 1,461 years to make a full cycle back to where it started. By making a correction of 13 days every 52 years, they could have avoided this. I believe it may also be the primary reason why the number 13 was considered so special by all the peoples of Central America: it held the key to reconciling their sacred calendar with the stars.

AGES OF THE SUN

For the Aztecs, the idea that the Sun would die if it was not fed was not some idle fantasy. They believed that this had happened four times before, and that each time mankind had been all but wiped out in some cataclysm. One version of this story is contained in an anonymous script called the *Leyenda de los Soles*, or Legend of the Suns.

The manuscript dates from 1558, more than 30 years after the conquest, and is written in Spanish. It contains the following details concerning supposed earlier ages, which it calls 'suns':

First Sun	*Nahui Ocelotl*	Duration: 676 years (52 x 13)
Second Sun	*Nahui Ehecatl*	Duration: 364 years (52 x 7)
Third Sun	*Nahui Quiahuitl*	Duration: 312 years (52 x 6)
Fourth Sun	*Nahui Atl*	Duration: 676 years (52 x 13)

Looking at the Aztec calendar stone itself, I could see evidence of how it recorded these past ages. Around the central disc was a symbol that is sometimes called ollin, or movement, by archaeologists. This symbol is not unique to this stone alone but is frequently shown on other Aztec calendars, including one in the British Museum.

With a shape that is possibly based on the appearance of a serpent's face when viewed straight on, it consists of four squares arranged asymmetrically around the centre. At either side and joined to these squares are loops containing the 'hands' with eagle's claws crushing hearts. The four squares are filled with glyphs representing the days on which the previous ages came to an end. I could see that the first of these, Nahui Ocelotl, is represented by an ocelot or jaguar head with four dots, i.e. the day named 4 Ocelotl in the Tonalamatl cycle of 260 days. According to Aztec mythology, this was an age of giants. It was brought to its end when they were attacked and destroyed by jaguars. The symbol was shown in the top right-hand square of the ollin sign, close to the glyph of an obsidian dagger, or *tecpatl*, representing the direction of east.

Looking at the stone, it was clear that the cycle of ages went anticlockwise. This at first seemed odd until I remembered that, although the Sun appears to rise in the east and set in the west, this is only because of the rotation of the Earth. The Earth's annual orbit around the Sun makes the latter appear to move through the zodiac in an anticlockwise direction. Thus the symbol for the second age, 4 *Ehecatl*, was to be seen in the top left-hand corner of the stone. Ehecatl was an aspect of the god Quetzalcoatl. He was evidently also associated with the north, as adjoining the square containing his symbol was a warrior's headdress symbolising that direction.

The next square in the anticlockwise sequence was in the bottom left of the ollin symbol. It contained the glyph 4 Quiahuitl ('4 Rain'), that was partly composed of a head representing the rain god Tlaloc. As was to be expected, it was placed close to a symbol representing Tlaloc's house in the west. Tlaloc, whose masks alternate with those of Quetzalcoatl on the little pyramid at the heart of the 'citadel' at Teotihuacan was the god of rain – rains of fire as well as water. The age over which he ruled was said to have been brought to its end by a rain of volcanic lava and ash.

The last square, in the lower right-hand corner and close to the symbol for south, contained the glyph 4 *Atl*, or 4 Water, conjoined with a stylised head representing the water goddess Chalchiuhtlique. Accordingly, this age was brought to an end with a flood.

The theory behind the Aztecs' ages of the sun gets a little

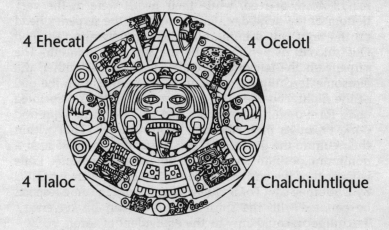

4 Ehecatl

4 Ocelotl

4 Tlaloc

4 Chalchiuhtlique

Figure 7: The centre of the wheel

tricky to understand. However, as can be seen, the number 52 features prominently in this account. It is clear that if the durations of the second and third 'suns' are added together, the sum comes to 676 (or 52 x 13) years: the same as Nahui Ocelotl and Nahui Atl. This implies that, as given, the table of ages is actually only three-quarters of a full cycle, which should comprise 52 x 52 years, i.e. 2,704 years in all. The explanation for this discrepancy is that the Aztecs believed there would be five ages in all: our own being the fifth and last. This last age is represented on the calendar stone by the shape surrounding the eagle claws of Tonatiuh and the symbols of the four previous ages. Taken with the four round dots (above and below the eagle claws), this gives the glyph '4 Ollin'. This means '4 Movement' and it symbolises the fifth and last age. This should have a notional period of 13 x 52 years, i.e. 676 years and like the first age is ruled over by the ocelotl, or jaguar.

Around the very outer circumference of the calendar stone, I could make out the forms of two enormous *Xiuhcoatls*, or fire serpents. The bodies of these serpents were composed of scales – eleven on each – bearing the symbol for fire. The tails of the two serpents met at the top, either side of a symbol that apparently represented the day-sign for 13 *Acatl*, or reed, while their heads were at the very bottom of the huge disc. Emerging from the serpent's head on the right was a human-like head representing the god Quetzalcoatl as personified by the sun god Tonatiuh. The serpent on the left disgorged the head of his brother, the fearsome Tezcatlipoca, personified as Xiuhtecuhtli, the god of the night. Together, these serpents, with their associated gods, form a pair of opposites, like yin and yang, light and dark, or, at its most abstract, good and evil. That within their culture the god of darkness, Tezcatlipoca, held such a dominant position says much about Aztec culture. Little wonder that the early Christian missionaries were appalled by what they saw. For if Quetzalcoatl could in some respects be equated with the spirit of Christ, then his adversary, Tezcatlipoca, could only be the equivalent of Satan.

The *Leyenda de los Soles* is not the only colonial source documenting Aztec beliefs concerning world ages. Another, and in many ways more interesting, document is the so-called *Vatico Latin Codex*. According to this document, the four ages were as follows:

> The First Sun was called *Matlactili*. It had a duration of 4,008 [years] and the people who lived then were maize-eating giants. At the end of this age, the Sun was destroyed by a flood. Fortunately, some men were turned into fish and so survived the flood. According to some accounts, only one couple survived (*Nene* and *Tata*), while others say there were seven couples who lived in a cave. They afterwards repopulated the Earth. This era was presided over by the goddess of water, Chalchiuhtlique, who was the wife of Tlaloc.
>
> The Second Sun was called Ehecatl. It had a duration of 4,010 years. During this time, people ate a wild fruit called *Acotzintli*. This sun was destroyed by Ehecatl, god of winds. Some people survived by turning into monkeys and clinging onto trees. The destruction happened in a year called *Ce Itzcuintli* (One Dog). One man and one woman, who stood on a rock, were saved from destruction. This period was a golden age and was presided over by the god of wind.
>
> The Third Sun was called *Tleyquiyahuillo*. It had a

duration of 4,081 years. The men living then were descendants of the couple who survived the Second Sun. They ate a fruit called *Tzincoacoc*. This world was destroyed by fire on a day named *Chicunahui Ollin*. This age was given the name *Tzonchichiltic* ('Red Head') and was presided over by the god of fire.

The Fourth Sun was called *Tzontlilac*. It began 5,026 years ago. This age was called Tzontlilac ('Black Hair') and it was during this time that Tula was founded. Men died of starvation after a rain of blood and fire.

This account is quite different from that described in the *Leyenda de los Soles* but is perhaps more authentic. It doesn't say who the god ruling over the third age was, but we can make our own inferences. Rather as Christians think of the Holy Trinity as three aspects of the same unitary God, so according to Aztec mythology the primary god exists as a divine couple. This god, *Ometeotl*, exists in the 13th and highest heaven, which they called Omeyocan, or place of duality. His/Her divine aspects were the twin principles of masculine and feminine that they called *Tonacatecuhtli* and *Tonacacihuatl*, or Lord and Lady of our Sustenance. From this divine couple were born all the other, lesser gods beginning with four sons called in order Red Tezcatlipoca, Black Tezcatlipoca, Quetzalcoatl (one aspect of whom is Ehecatl the god of wind) and *Huitzilopochtli*, the primary patron of the Aztecs. Given the descriptions of 'red' and 'black' heads, we can infer that the ruler of the third age was Red Tezcatlipoca and of the fourth Black Tezcatlipoca. The subtleties of these differences are lost on us today. However, the Black Tezcatlipoca is the one who features prominently in Aztec mythology as the rival of Quetzalcoatl. He was greatly feared and, along with Huitzilopochtli, required frequent human sacrifices if his anger were to be averted.

The duration of these ages, which is strikingly different from the periods given in the *Leyenda* account, seems arbitrary and at first difficult to understand. Internal evidence of the manuscript suggests that it was written in 1576 or thereabouts. This would mean that, according to this manuscript, the fourth age began in *c.* 3450 BC. From this, we can infer that the third age began in *c.* 7561 BC, the second in *c.* 11541 BC and the first in 15549 BC. What, if anything, these figures mean in practice is hard to say.

What is clear from these accounts is that the Aztecs thought in terms of cycles which followed the pattern of seasons and compass direction. An entire circulation of the ages depicted on the Aztec calendar stone was the equivalent, but on a very much longer timescale, of a day. The Aztecs, however, were only one of the nations inhabiting Central America, and it would appear that their knowledge of time-cycles was limited. The real masters of time were the Maya, who, unlike the Aztecs, were able to write using a syllabic alphabet. As a result, although their civilisation reached its height centuries before the advent of the Aztecs, they have left us far more information about beliefs that were once held in common throughout Central America.

CHAPTER 4

The Rediscovery of the Maya

In the sixteenth century, Catholic friars, such as Bernadino Sahagun, were responsible for both destroying and preserving Aztec knowledge. Meanwhile, as Spanish control spread to Guatemala and other areas of Central America, so remnants of other ancient civilisations began to come to light. These relics of a lost world had been swallowed up by the all-encompassing jungle centuries before the arrival of Cortés. They were not easily reached by the Spanish, and as a result for the most part escaped the sort of irreparable damage done to the archaeological record in the Mexico Valley area. In the seventeenth and eighteenth centuries, as Europeans began to explore the jungle areas, they began to stumble over these ruins, most of which were Mayan in origin. However, the story of Mayan archaeology really begins in 1773 with the discovery of Palenque, a till then unknown city, by a Father Ordoñez of the cathedral town of Ciudad Real in Chiapas.

This extraordinary city, more and more of which is being unearthed each year, is probably the most beautiful in all of Central America. Its pyramids and temples, which are built out of white limestone, are not only well crafted but reveal an aesthetic taste on a par with Renaissance Europe. Many of these monuments carry large entablatures, some of which depict what are clearly either mythic or historic scenes. Others are covered with what have long been recognised as hieroglyphs, which till recently remained untranslatable.

Who had built Palenque and when became a matter of burning curiosity that has occupied scholars ever since Father Ordoñez wrote about his findings in a monograph entitled *A History of the Creation of Heaven and Earth.* He explained the fabulous ruins he saw as the work of non-indigenous people who came from across the Atlantic. They were led, he said, by a man called Votan from the city of Tripoli. Votan, evidently, made several trips back home and also visited a city where a temple was being built that would reach up to heaven. Good Christian that he was, Father Ordoñez identified this with biblical Babylon. This made sense at the time as the pyramid temples of Palenque are, in some respects, similar to the ziggurats that can still be seen in Mesopotamia. It therefore followed that Votan, or maybe some earlier seafarer, had brought the idea of ziggurat building across the Atlantic and into Central America. Thus could be explained the relatively advanced civilisation of Mexico relative to the rest of North America: it was an import from the Old World.

Father Ordoñez claimed to have found the story of Votan in a Quiché Maya book which had been partially copied by Bishop Nuñes de la Vega prior to the burning by him of the original text in 1691. According to the book, Votan had arrived with a party of other men, all of them wearing long white gowns. They had evidently been well received by the natives of the area, who submitted to his rule, and they had subsequently married their daughters. This implied that the dynasty who built Palenque (and perhaps other cities in the area) was of mixed race: partly indigenous and partly African or Asiatic.

Whether there is any truth in this legend is still a matter of heated dispute, especially as, like so many other important documents, the book in which Ordoñez claimed to have found the story has since disappeared. However, it is worth noting that when Cortés arrived on the Island of Cozumel in 1519, he received word that there were two Spaniards living among the natives. They were the only survivors of a previous expedition to the island that set out in 1511. One of them we have already heard about: Geronimo de Aguilar, who subsequently joined Cortés's expedition and provided invaluable service as an interpreter. The other, Gonzalo Guerrero, went native. Diego de Landa (and other early chroniclers) records that he adopted Indian manners, grew his hair long, had his body tattooed, pierced his ears and followed their idolatrous religion. He married a chief's daughter and was so well thought of by her tribe that he was put in charge of their military affairs.

Like Aguilar, Guerrero could have joined Cortés's expedition and gone back to the life of a Spanish conquistador. However, he preferred his new life over his old and stayed with the Indian tribe of which he was now an important nobleman. If such a 'reverse conversion' could happen in the sixteenth century, might not the same thing have happened with a Phoenician mariner called Votan? At least Guerrero had the opportunity of going back to his old life. If there really was a Votan, then it is likely he was a Phoenician and that his ship was blown off course by a gale in the Atlantic. If it was wrecked on the coast near Palenque, then it would probably have been impossible for him to return to Tripoli or anywhere else in the Old World. Like it or not, he and his men would have had to make the best of their situation. Under the circumstances, marriage to a chief's daughter and with it a leading role in Palenque would have been a very satisfactory outcome. It would certainly have been better than Aguilar's fate of being enslaved. Thus, we should not, in my opinion, be too hasty in dismissing the story reportedly copied by Bishop de la Vega. As will be discussed later, it just may be based on truth.

Publication of Ordoñez's own book, *A History of the Creation of Heaven and Earth*, provoked a wave of interest in Mayan monuments in general and Palenque in particular. Now that Catholicism was firmly entrenched throughout Central and Southern America, relics of the old religion were no longer regarded as such a threat. Whereas before they were mostly destroyed on sight, they could now be studied objectively as works of art.

The first person to make a proper survey of the ruins of

Palenque was an army captain called Don Antonio del Rio. In 1787, he employed teams of local Mayans to cut back the jungle, thereby revealing one extraordinary building after another. Drawings of the major buildings and plaster-casts of their stucco reliefs were made. In his report, which he submitted to the authorities in Madrid, del Rio proposed that some of the buildings may have been of Roman manufacture, in particular an aqueduct running under the largest building, now called 'The Palace'. To put this apparently extraordinary suggestion into context, he quoted the authority of a Dominican friar named Jacito Garrido. He had claimed that North America had been visited in antiquity by people from Greece, Britain and other places. The implication was that the Romans were not the only, or necessarily even the first, Europeans to navigate the Atlantic.

In Madrid, del Rio's report was simply filed and it would have probably been lost altogether had not a copy of it remained in Guatemala City. Following the end of Spanish rule in the wake of the Napoleonic Wars, this subsequently came into the hands of an Italian: Dr Felix Cabrera. He wrote his own foreword to del Rio's report in which he suggested that America may have been first visited by Atlas, the eponymous father of the Atlanteans, then by Hercules and finally, just prior to the First Punic War (264 BC), by the Carthaginians. This amended report eventually found its way into the hands of a London publisher: Henry Berthoud. He added to it a collection of plates and in 1822 published del Rio's report, along with Cabrera's addition in book form under the title *Description of the Ruins of an Ancient City discovered near Palenque.*

The publication of this book caused a storm of interest in all things Mayan, and very soon other, more serious investigators, such as John Stephens, Alfred Maudsley and Claude Charnay, were visiting Palenque and other Mayan sites. The first and in many ways most important of these individuals was Charles Étienne Brasseur de Bourbourg. He was born in 1814 and ordained in 1845. For a short while, he was professor of ecclesiastical history at Quebec before finding his true vocation: ferreting among the archives of church libraries. In 1848, after a brief return to Europe, he moved to Mexico City, where he took on the post of chaplain to the French legation. During his time in Central America, where he lived off and on for the next 17 years, he combined teaching with exploration. Anxious to learn as much as possible about the ethnology and history of the Native Americans, he searched the libraries of both Europe and the newly emergent countries of Central America for ancient manuscripts. This work was not without its fruit.

Brasseur made his first important find while searching a library in Guatemala. Here he found a book that was written in the Quiché Mayan language using the Roman alphabet shortly after the conquest of Guatemala. It had been preserved by the Mayan elders of the city of Quiché, now called Chichicastenango. Around 1702, this book was shown to a Dominican friar called Francisco Ximenez. He made the only known surviving copy of the work, adding a Spanish translation alongside the Quiché original. Following the closure of the monasteries in 1830, Ximenez's work, which till then had been kept under wraps in his Dominican monastery, found its way to a library in Guatemala City. It was this book, today known as the *Popol Vuh*, or Book of Council, that Brasseur came across while resident there. He took it back to Paris and the Quiché text was published alongside his own French translation.

The *Popol Vuh* was not the only important work to be rediscovered by this enterprising cleric. Another manuscript copied by Ximenez called *The Annals of the Cakchiquels* also came into his hands. The Cakchiquels, who live by Lake Atitlan, are closely related to the Quichés. This book is important because, although different from the better known *Popol Vuh*, it confirms many of the historical events recorded in that text. Back in Europe, Brasseur made further important discoveries. In a library in Madrid, he found the manuscript of the *Relación de Yucatán*. This book, written by Friar Diego de Landa, the first Bishop of Yucatán, contains what has turned out to be a treasure beyond belief: the key to successfully translating the Mayan hieroglyphs.

While in Madrid, Brasseur also met a descendant of Cortés himself: Jean de Tro y Ortalano. Ortalano had in his possession part of one of the very few surviving Mayan bark-books from before the conquest. Called the *Troano Codex*, it has now been reunited with its other half (the *Cortésianus Codex*) to make one large work: the *Madrid Codex*.

Reading through his assembled collection of manuscripts, which included copies of many other rare documents, Brasseur could not help but notice that the Mayans claimed that the ancestors of the modern-day Maya, as well as the Yaqui, or peoples of northern Mexico, claimed to have come from a city called *Tulan*, 'Place of Reeds'. Prior to this, they had lived in a land somewhere over the sea. Not unnaturally, Brasseur assumed that this land must have been one and the same as Plato's lost continent of Atlantis, which he believed extended all the way from the West Indies to the Canary islands. The supposition that the Maya came originally from Atlantis was no doubt strengthened by the fact that so many of the place-names of ancient Mexico, Guatemala and El Salvador were based on variants of the root '-titlan', e.g. Lake *Atitlan* in Guatemala, *Utitlan* in Mexico, *Cuzatlan* in El Salvador and, of course, *Aztlan*, the supposed original island home of the Aztecs. To Brasseur, and to other, subsequent 'Atlantophiles', it was obvious that these name-endings were derived from the name of the mother-city of Central American culture: Atlantis itself.

Brasseur recorded his findings in a large work entitled *Histoire des Nations Civilisées du Mexique et de l'Amérique Centrale*, which was published between 1857 and 1859. Though these days, his recovery of ancient manuscripts notwithstanding, he tends to be regarded as a crank because of his belief in the reality of Atlantis, in the nineteenth century he was very highly thought of throughout Europe. Indeed, he was considered a Maya expert. It is therefore not surprising that he and another remarkable Frenchman, Count Jean Frederick Waldeck, were asked to collaborate on another book – this one commissioned by the Emperor Napoleon III himself – which was subsequently published in 1866 under the title *Monuments anciens du Mexique, Palenque, et autres ruines de l'ancienne civilisation*.

Like his co-author Brasseur, Waldeck was an adventurer who loved exploring Mayan ruins. An artist rather than a serious scholar, he was the first Westerner to make a serious attempt at drawing Mayan ruins. Very much in the mould of one of the first Napoleon's *savants*, he spent several years studying and making sketches of the ruins at pre-Columbian sites. During his time at Palenque, he lived in one of the Mayan buildings which for this reason is still known today as the House of the Count. Like Brasseur, and indeed most other early investigators, Waldeck was a 'diffusionist'. He believed the origins of Mesoamerican civilisation were to be found not in the Americas but overseas. In particular, he was convinced that the Mayans learnt the art of pyramid building from the Egyptians. Among many other projects he worked on were the original illustrations for Lord Kingsborough's mammoth work entitled *Antiquities of Mexico*. This work, somewhat similar in concept to the famous *Description de l'Egypte* composed by Napoleon Bonaparte's *savants*, contained many original drawings and reproductions of scarce manuscripts. Unfortunately, Waldeck's attractive illustrations were not true to the original sketches on which they were based. To the annoyance of later scholars, he placed artistic licence over accuracy.

The first photographs of Mayan ruins were taken by another Frenchman, Claude Charnay. He too was sponsored by Napoleon III, this time under the auspices of Viollet-le-Duc, minister for fine arts. Something of a pioneer in field photography, Charnay took pictures of the Pyramids of Teotihuacan. At Palenque, he not only took pictures but made papier mâché casts of some of the reliefs, all of which he sent back to Paris. Like Brasseur and Waldeck, Charnay was a diffusionist. However, his ideas on diffusion came from quite a different direction as, noting similarities in dress, pyramid building and language, he proposed that the culture of Central America originated in the Far East.

Charnay was not the only nineteenth-century Frenchman to take photographs of Mayan ruins. Around the same time as he was working, Augustus le Plongeon (a native of Jersey and therefore, strictly speaking, British) began investigating the ruins of Yucatán. A maverick even by the standards of early Mayanists, le Plongeon was a diffusionist with a difference. Far from accepting, as most people then did, that civilisation had been brought to Central America from overseas, he believed that the cultural traffic had gone in the opposite direction. Basing his arguments on similarities of building techniques, styles of clothing and ornamentation, he proposed that people had migrated from Mexico to the Far East. From there they had travelled westwards, eventually arriving in Egypt, where they again built pyramids.

Ignoring the obvious objection that the Egyptian civilisation preceded the Mayan by millennia, he set off for Mexico to prove that he was right. He and his wife Alice, also an accomplished photographer, arrived at Chichen Itza around 1874. There they stayed for some months, taking photographs using a 3D camera and carrying out the first systematic survey of the ruins. Le Plongeon had one advantage over earlier explorers in that he was an accomplished linguist. He learnt sufficient Yucatecan Mayan to be able to converse with the local people, most of whom did not speak anything else. They told him where to dig if he wanted to find something remarkable. They guided him to a major discovery: the first known example of a 'Chac-Mool' statue. It was buried at a depth of 24 ft near to the Platform of Venus, and had presumably been put there by the Spanish at the time they were seeking to rid the region of its old religion. Today, this statue, which is of a semi-reclining man holding an offerings plate over his stomach area and looking over one shoulder, is in the Museum of Archaeology in Mexico City as one of their most prized items.

ORTHODOXY SETS IN

The French, Germans, Italians and Spanish were not the only nineteenth-century Europeans to take a keen interest in Mexican antiquities: the British too were involved. This interest began in earnest with the publication in 1841 of a travel diary in two volumes entitled *Incidents of Travel in Central America, Chiapas and Yucatán* by the American author John L. Stephens. This was followed two years later by a sequel: *Incidents of Travel in Yucatán*.

Humorous and filled with anecdotes, these books appealed greatly to armchair travellers and were huge bestsellers. However, though Stephens's style of writing was light – in many ways he was the Bill Bryson of his day – his attitude

Figure 8: John Stephens

towards his work was much more serious. He and his travelling companion, a British artist named Frederick Catherwood, spent months in the field studying and recording what they saw. They visited numerous sites, not just the obvious ones such as Chichen Itza, Uxmal and Palenque but also Copan, which had seldom before been visited by Westerners. They carried out meticulous surveys at all the ancient sites they visited. Their reports were greatly enhanced by the extraordinarily detailed drawings made by Catherwood that were later used for making lithographic plates to illustrate Stephens's books. As many of the structures they viewed have deteriorated considerably in the century and a half since they were writing, Catherwood's illustrations are still of immense value to archaeologists working in the field today.

Throughout the latter part of the nineteenth century, British and American interest in Central American antiquities was growing ever stronger. During this period, the most important British contributor to Mayan studies was a classics scholar and former consul general to Tonga: Alfred Maudsley. Like Le Plongeon, whom he knew and sometimes collaborated with, Maudsley was a great believer in the importance of photography as a primary means of recording artefacts in situ. Unlike him, he was a scientist and therefore not inclined to pay too much heed to romantic notions of Egyptians (or even 'Lost Tribes of Israel') migrating to Central America to build pyramids. Maudsley worked hard, visiting sites throughout the jungle areas of Peten and Chiapas. Using a large-format, wet-plate camera, he took pictures of everything important that he saw. Where possible, he also took plaster-casts of stelae and mouldings. These he knew would be of great importance to scholars who would not be able to see the real things.

Returning back to London, he commissioned a skilled artist, Annie Hunter, to create detailed lithographic plates

Figure 9: Catherwood's drawing, 'General view of Palenque'

from his photographs and casts. His work was eventually published in 1889 as an addendum to a multi-volume work entitled *Biologia Centrali-Americana*. In all, it comprised five volumes, one of text and four of accompanying plates. This was the most important collection of Mayan hieroglyphs to date. Because it was so accurate, it is still of use to scholars today, not least because many of the monuments depicted have since either disappeared or sustained damage.

Maudsley's work, which was studied avidly in Europe and America, ushered in a new era. The days of the self-financing amateur were coming to a close. The next generation of researchers would be looking for financial backers from the academic sector. Two major institutions emerged as patrons of what was now becoming a respectable branch of archaeology: the Carnegie Institution of Washington DC and the Peabody Museum of Harvard University. The former became the dominant force in the early part of the twentieth century by virtue of the work of their protégé: Sylvanus Griswold Morley, who became director of the Carnegie Archaeological Program in the Maya area from 1914 to 1929.

Morley's main concern – indeed obsession – was in recording and, if possible, translating Mayan hieroglyphs. Though in the main this was to prove an impossibility for another 50 years or so, this did not deter him from publishing in 1915 the first student textbook on the subject: *An Introduction to the study of Maya Hieroglyphs*. Given that the vast majority of inscriptions (other than dates) were at the time untranslatable, this seems an impertinent title. However, Morley was an accepted expert on the complexities of the Mayan calendars and their methods for inscribing dates on their monuments. His book covers these subjects in extraordinary detail, drawing on his own vast collection of illustrations made on numerous field-trips to remote locations.

Morley's work was a watershed in other ways, too. His systematisation of the study of date hieroglyphics paved the way for a more rigorous and scientific analysis of Mayan monuments. No longer would it be possible or acceptable – at least among the new professionals – for a Mayanologist to talk glibly about either Atlantis or

Egyptian migrations. The accurate dating of monuments from their inscriptions allowed the creation of an accepted chronology for Mesoamerican civilisation that was independent of native records, on which doubt had been cast, such as the *Popol Vuh, The Annals of the Cakchiquels* or the *Books of Chilam Balam*. Theories of Mayan origins based around diffusionism were now replaced by one of independent evolution. It was now taught that the Aztec and Mayan civilisations were the latest flowerings of a self-created, self-evolving Mesoamerican culture. Discussion of anything else – most of all diffusionism – was tantamount to heresy.

The doyen of this mid-twentieth-century attitude towards the Maya was another British archaeologist: J. Eric Thompson. A student and disciple of Morley, he worked with him for many years at Chichen Itza and other sites throughout the Yucatán, Chiapas and Peten. When Morley retired, Thompson, then a fellow member of the Division of Archaeology at the Carnegie Institution, inherited his mantle as senior practising Mayanologist. However, he left the Carnegie (which was soon to pull out of Mayan studies altogether) shortly afterwards. Instead, at the end of 1926, he took up a post with Chicago's Field Museum of Natural History. This was a situation that suited him admirably, as he was given virtually free rein to do as he pleased. In the decades that followed, for political reasons, Mexico was to become a virtual no-go area for American archaeologists. Thompson, however, was able to redirect his energies towards Belize. The fact that this was then the Crown Colony of British Honduras and he was himself a British subject made the obtaining of permissions to dig that much easier.

Thompson was a prolific writer who wrote more than 24 major articles and books between 1929 and his death in 1975. In these, he solidified the notion that Central American civilisation was self-produced. Because of the almost unique authority that he exercised over his contemporaries, who nearly all seem to have been intimidated by his undoubted intellect, this idea moved from being a theory based on observations to a dogma. It became impossible for any self-respecting archaeologist to suggest that Mesoamerican culture was a transatlantic or transpacific import, still less that it originated in Atlantis.

In his book *The Rise and Fall of Maya Civilization*, Thompson presented an antidote to the 'disconnected' atomistic approach to Maya studies. Basing his account on carbon-14 dates as well as decoded inscriptions, he laid out a chronology that is still the basis of Maya studies today. This begins with a 'Formative Period' (*c.* 1500 BC–AD 200), followed by a 'Classic Period' (*c.* AD 200–AD 925). This 'Classic Period' subdivides into three shorter periods: the 'Early' (AD 200–AD 625), 'Florescence' (AD 625–AD 800) and 'Collapse' (AD 800–AD 925). Thompson's 'Classic Period' is followed by a 'Mexican Period' (AD 925–AD 1200), when the Mayalands were supposedly invaded by Toltecs from the Valley of Mexico, who either enslaved the indigenous Maya or took over ruling positions. His final period is that of 'Mexican absorption' (AD 1200–AD 1540), during which new 'Mayan' empires grew up in place of the Toltec hegemonies: Mayapan in Yucatán; the Quiché in the Guatemalan highlands. This last period culminates, of course, with the Spanish conquest of Mayaland that brought the last vestiges of these empires to an end.

This grand scenario of history, probably implicit in the writings and work of earlier authors, was Thompson's greatest contribution. With variations, it is still used as a basic framework today. However, these days the 'Formative Period' is referred to as the 'Preclassic' and divided into three sub-periods: Early Preclassic (1800 BC–1000 BC), Middle Preclassic (1000 BC–300 BC) and Late Preclassic (300 BC–AD 250). The 'Mexican Period' is termed the 'Early Postclassic' and the 'Mexican absorption' period is called the 'Late Postclassic'. To these is added another period, the 'Archaic', that stretches indeterminately back before 2000 BC – perhaps as far back as the end of the Pleistocene or last ice age – during which time there was no recognisable civilisation as such.

That Thompson's broader picture has stood the test of time so well is testimony to its usefulness in defining archaeologically meaningful periods. However, that does not mean that it necessarily gives a correct view of history. In particular, it does not mesh especially well with the written evidence of such Mayan writings as the *Popol Vuh* or Aztec codices such as *Leyenda de los Soles* that talk in terms of ages prior to our own when mankind was all but destroyed by the gods. Nor does the theory that Mesoamerican civilisation evolved independently from the rest of the world fit well with stories told by both Aztecs and Maya of a migration from a land beyond the sea. Such stories had, of course, inspired earlier writers to entertain ideas of diffusion, either from the lost continent of Atlantis, Africa, Europe or Asia.

Thompson dismissed such suggestions out of hand, writing: 'No specialist in the field supposes that America was populated by immigrants from across the Atlantic or from across the Pacific . . .'.[1] He, like most of his generation, assumed that the ancestors of all native Americans were Asiatics who came to the Americas via a land bridge at some time around 11,000 BC:

> Archaeologists agree that America was populated by immigrants from Asia who crossed the Bering Strait; there is less unanimity as to when the earliest crossings were made, majority opinion favoring about twenty thousand years ago . . . The earliest evidence of man (at Tule Springs, Nevada) carries us back to 11,000 BC, but more ancient sites will certainly turn up.[2]

Thompson's prophecy was right: more ancient sites have turned up. However, he would have been perplexed to discover that the oldest of these is not in Alaska, close to where the supposed early migrations over a land bridge would have taken place, but rather in Brazil. Here, in a rock shelter at Pedra Furada in the Capivara National Park, remains of human occupation have been found that go back to at least 50,000 BC. Drawn on the walls of the rock shelter are figures of hunters (and the animals they preyed upon) not dissimilar to examples to be found in Africa and Europe. Even a cursory look at a map reveals that Brazil is much nearer to West Africa than Alaska. True, to get from Africa to South America would have meant navigating the Atlantic. However, there are favourable currents and trade winds that would have made this relatively easy if these ancient people had boats of some sort. The Capivara National Park is in Piaui Province and its location suggests that people could have reached the rock shelter after navigating up one of the lesser rivers of Brazil, such as the São Francisco.

In recent years, as we shall see, it has become possible to read the Maya hieroglyphs. This work is still ongoing but already it has caused a profound change in attitude towards the Maya. Gone now is the romantic belief that theirs was a pacifistic way of life, intent only on raising crops, writing books and building cities. We now know that, like other native American peoples, they fought frequent wars with one another. To our horror, it seems that one of their principal war aims was a perceived need for human sacrifice. Rather than killing each other in battle, war was a ritual conducted with the aim of capturing neighbouring lords. If they were fortunate, these nobles might be sacrificed to the gods straight away. More often, they were kept as slaves and ritually bled from time to time. Why these otherwise quite civilised peoples should indulge in such macabre practices is something that is difficult to explain but it seems that, like the Aztecs, they feared something terrible would happen to the world if they didn't.

What we also know about the Maya is that they were nowhere near as materialistic as we are. All that most people owned was the clothes they stood up in and perhaps a few household utensils, weapons or tools. A Stone Age people right up until the time of the Spanish conquest, they cultivated the soil using simple stone or wood implements to grow such staples as maize, beans and sweet peppers. If their monuments are anything to go by, their richly attired rulers were different. When they weren't fighting wars, and possibly even then, they wore elaborate robes. These, however, were not personalised fashion statements but ceremonial costumes. Translation of the hieroglyphs reveals that their headdresses, bracelets and skirts had symbolic significance. Like the vestments of a Roman Catholic priest saying mass, these robes had a meaning that was altogether religious to its core.

Because they did not possess metals, did not use wheels and

had little knowledge of navigation, their Dark Age contemporaries from Europe, China or the Middle East would undoubtedly have regarded the Mayan civilisation as primitive. However, in one important respect they were far in advance of any of them: astronomy – more specifically, the awareness of time-cycles, long and short, as they relate to the movements of the Moon, the planets and the Sun itself. This was tied in with their most important religious belief: that we are living in the fourth age of the Sun. Like the Aztecs, the Maya believed that there had been other 'suns' before our own and that these ages had been characterised not only by the rulership of different gods but the presence on Earth of different types of mankind.[3] In each case, these races were destroyed by cataclysms that left few survivors.

THE MAYAN INHERITANCE

Contrary to popular belief, the Maya as a people did not die out with the decline of their civilisation. Although nearly all their 'Classic' cities, for reasons which are still not clear, were abandoned to the jungle by around AD 900, they themselves survived. Today there are more than six million pure-blooded Maya, still living in the same region of the world once inhabited by their ancient ancestors. This region encompasses the south-east of Mexico (principally the Yucatán Peninsula but also Chiapas and part of Tabasco Province), the whole of Guatemala and Belize (formerly British Honduras) plus the most westerly parts of Honduras and El Salvador. Though divided politically and living in different countries, the Mayans in this fairly large region of Central America are all one people. And, prior to the Spanish invasions of the sixteenth century, they all practised the same religion, albeit with local variations.

Regional variations extend also to language. The Maya of today (as in the past) speak a wide variety of languages. Just as the European languages that are spoken throughout the Western world have a common origin in an Indo-European proto-language, so the languages of the Maya all stem from the same root-language: what archaeologists refer to as 'proto-Mayan'. Actually, Mayan languages are more closely related to each other than, say, German is to Greek. They are not always mutually unintelligible. Just as European languages are grouped into 'families' (e.g. the 'Romance' languages of France, Spain and Italy, the 'Germanic' languages of northern Europe and the 'Celtic' languages of Brittany, Wales, Scotland and Ireland), so the Mayan languages form similar groups.

Thus, while there are over thirty Mayan languages spoken today, they belong to just six major language groups: Huastecan, Yucatecan, Greater Cholan, Greater Manjobalan, Mamean and Greater Quichean. Of these, the most important today (because the most widely spoken) are the Yucatecan and Quichean families. However, in the past, Cholan Mayan was much more widely spoken than it is today. The region occupied by Mayans speaking Cholan once extended right through the centre of Mayaland, from the coastal region of Tabasco, along the valley of the Usumacinta river, to Belize. This is the region where the great Classic cities of the Maya were built: places like Tikal, Yaxchilan, Bonampak and Palenque. Archaeologists are therefore certain that this was the language spoken at the time. It is undoubtedly the one in which the many inscriptions recorded on the monuments of this area were written.

Today, a major centre of speakers of Tzotzil and Tzeltal – Cholan languages – is the region around the city of San Cristobal de las Casas. The people who live here are among the most conservative of all Mayans and most of their women continue to wear national dress: the distinctive *huipil*, or embroidered blouse. Despite centuries of exposure to Christianity and an outward adherence to Catholicism, they have preserved elements of the old culture. Almost uniquely among the Maya, the Tzotzils preserved the old 260-day calendar of Mesoamerica which the Aztecs called the Tonalamatl and the Maya refer to as the *Tzolkin*, or count of days. Today, the Tzolkin is making something of a comeback, as many Maya and other indigenous peoples of America seek again the old ways. The old Tzolkin calendar is central to this revival as it is the basis of fortune-telling. Shaman priests, called 'day-keepers', use it for their prognostications.

Figure 10: Mayan day-names

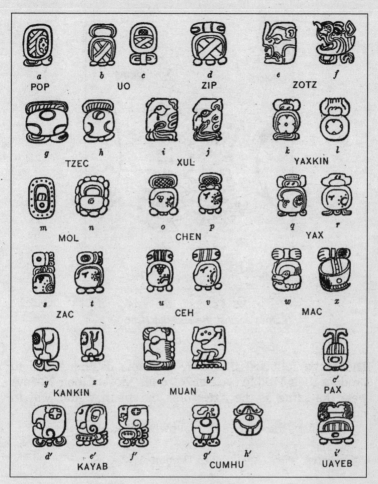

Figure 11: The month signs in the inscriptions

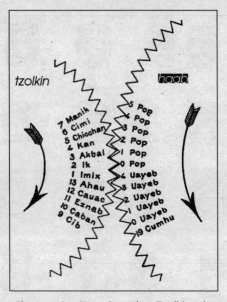

Figure 12: Counting the Tzolkin day-names against the 365-day Haab

The Maya followed the same principle as the Aztecs to produce their Tzolkin cycle of 260 days. Most of their 20 days were the same as the Aztec ones except they had Mayan names.

In English, these have the following meanings:

Imix	Sea Dragon
Ik	Air
Akbal	Night
Kan	Corn
Chicchan	Serpent
Cimi	Death
Manik	Deer
Lamat	Rabbit
Muluc	Rain
Oc	Dog
Chuen	Monkey
Eb	Broom

Ben	Reed
Ix	Jaguar
Men	Eagle
Cib	Owl/Vulture
Caban	Force/Earth
Eznab	Flint/Knife
Cauac	Storm
Ahau	Lord

Like the Aztecs, the Maya also had a 365-day calendar, the *Haab*, composed of 18 months of 20 days' duration plus a 19th of only 5 days.

The 18 standard-length months are: *Pop, Uo, Zip, Zotz, Tzec, Xul, Yaxkin, Mol, Chen, Yax, Zac, Ceh, Mac, Kankin, Muan, Pax, Kayab* and *Cumku* (also spelt *Cumhu*). The short month of five days, considered very unlucky, is called *Uayeb* by the Maya. These months are represented by the Mayan glyphs in Figure 11.

Like the Aztecs, the Maya combined the two calendars, the Tzolkin and the Haab, to produce a calendar-round of 52 years.

However, they seem to have understood better the limitations of this system for counting long periods of time. Uniquely, the Maya date-stamped the monuments that they built. As we shall see, they did this by using a calendrical system that was based on the counting of days in relation to fixed events.

In terms of technology, the Mayan civilisation was undoubtedly primitive by our standards. They had no running water or electricity, no cars, computers, televisions, aircraft or even chariots. Yet, despite this material poverty, they were rich in other ways. Without the distractions of the modern world, they developed their psychic faculties in ways that we would find astounding. Like the Aborigines of Australia, they studied their dreams intently, aware that these had a profound effect upon daily life. To keep in touch with the 'other world' of gods and spirits, they developed a highly sophisticated system of blood-letting rituals that we would find repugnant. For, like ascetics everywhere, they believed that mortification of the flesh was necessary if life on Earth were to be maintained.

From early beginnings at the dawn of history, through a 'Classic Age' lasting from around AD 100 to 900, the Maya produced some of the world's greatest art. Then, for reasons which are still not fully understood, their great cities were abandoned. The previously most important regions of Mayaland, Chiapas and Peten, which lie in the central zone, were abandoned and reverted to jungle. For a few centuries more, in the northern region of Yucatán, a 'Post-Classic' Renaissance took place. This was a hybrid civilisation brought about by the merging of indigenous Mayan tribes with others from further south and with non-Mayans from the Mexico City area. The monuments they left behind, notably such baroque cities as Uxmal and Chichen Itza, are among the finest to be seen anywhere in Central America. This, however, could not hide the fact that the culture was in terminal decline, with human sacrifice becoming more and more commonplace.

Because of the vagaries of time, there is much that we will never know about the Mayan civilisation. However, the breaking of the code used by them in their hieroglyphic language has caused a revolution in our understanding. Discovering how this came about was the next step in my own journey.

CHAPTER 5

Decoding the Maya Legacy

With the possible exception of ancient Egypt, it is true to say that no field of archaeology has proved more challenging and ultimately rewarding than the rediscovery of the Mayan civilisation. Just as Egypt prior to the breaking of the hieroglyphic code by Champollion was a land of unknown mystery, so too up until very recently was the world of the Maya. It is only in the last 40 years or so that the Mayan hieroglyphs, a much more complex system than the Egyptian, have at last given up their secrets. Only now, as the present Mayan 'Age of the Jaguar' draws to its conclusion, are scholars able to read for themselves the many hundreds of inscriptions left by this enigmatic people and to begin to appreciate the full subtlety of this extraordinary culture. How this situation has come about is a story in itself and one that needs to be understood if we are to make a proper assessment of what has been discovered and, more importantly, of what this means in the larger context of world history and prophecy.

The bible for Mayan studies, at least in the nineteenth century, was the huge corpus of material that had been gathered together and published at great expense by Lord Edward Kingsborough. Running to nine volumes in all, each a massive tome, the *Antiquities of Mexico* was financially ruinous and its publication led to his early death in a debtors' prison. Lord Kingsborough's loss was scholarship's gain. His unwieldy tomes, copies of which are among the treasures of the British Museum and are today housed in its Library of Ethnology, contained all the most important sources of his time. These included the text of the *Popol Vuh* and Bishop Landa's *Relación de las Cosas de Yucatán* (both brought back to Europe by Abbé Brasseur de Bourbourg), as well as a copy of the *Dresden Codex*, the most important of the surviving Mayan bark-books.

The availability of such books, at least to those who had access to them, made it possible for a whole new class of armchair researchers to begin investigating the Maya. The first of these to begin the process of decipherment of the Mayan script in all seriousness was an eccentric gentleman named Constantine Samuel Rafinesque. Born of a French father and German mother in the suburbs of far-off Constantinople, he first visited America in 1802 and later made it his permanent residence. A self-taught naturalist, Rafinesque was well respected for his classification of numerous American plants – work carried out a generation before Charles Darwin shook the world with the publication of his *Origin of Species*. Categorising plants requires close attention to detail and the ability to recognise patterns among diversity. Rafinesque was good at this and able to apply these skills in other ways too. Keenly interested in the work of Champollion on the translation of Egyptian hieroglyphs, he decided to attempt what was then impossible: the translation of the Mayan. In this, he failed, partly because of the paucity of examples available to him. He did, however, make an important observation: that the Mayans used a dot and bar system of numerics. He also proposed that the language used in the hieroglyphs and in the written codices were one and the same but different from that of the Aztecs. In this, he was right, though it would be a long time before anyone would be able to read more than numbers on the Maya monuments.

As we have seen, the Maya and Aztecs both used 260-day and 365-day calendars, combining them to produce a longer cycle of 52 years. The discovery by Rafinesque of their dot and bar graphics now made it possible to read the

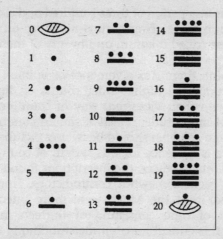

Figure 13: Bar and dot notation

numbers inscribed on monuments and thus to discover their most important calendar: the long count.

This revelation was again largely down to the patient work of armchair scholars. The first breakthrough was made by an American named Cyrus Thomas, who, armed with a partial copy of the *Dresden Codex* and Landa's *Relación*, deduced that Mayan numbers were to be read right to left and top to bottom, two rows at a time. Examining inscriptions of dot and bar numbers, Thomas saw that they worked in a vigesimal system, all of their individual numbers running from 0 to 19, i.e. from zero (represented by a cowrie shell) to three lines plus four dots (representing 19). He also realised that, like us, the Maya used a positional system to recognise higher-order numbers. However, there was a slight difference. Whereas we count in tens, hundreds, thousands, etc., because their system was vigesimal, they counted in twenties, four-hundreds, eight-thousands and so on, each order of magnitude being not ten but rather twenty times the one before.

The Mayans used their numerical system for business and commerce. However, for calendrical purposes, it had to be slightly adapted. In their calendar, a single day was called a *kin*, or sun. Twenty kins made up a *uinal*, or month. To give a reasonable approximation to a year, they counted eighteen uinals, a period of three hundred and sixty days, to make a *tun*, or stone. Twenty tuns made up an important period of time called a katun, and twenty of these made up a baktun.

At this point, there is a divergence of opinion as to how the Maya computed higher-order counts of days. If they carried on with their vigesimal way of counting, then the next order of magnitude, a *pictun*, should comprise twenty baktuns, and a *calabtun* should be twenty pictuns. However, though there is evidence for this system of counting in the *Dresden* and other codices, the inscriptions found on Mayan monuments seem somewhat contradictory. From these, it would appear that the Maya were more interested in a 'great cycle' of time consisting of thirteen, not twenty, baktuns. As it is critical for an understanding of the Mayan end-date prophecies, I shall return to this subject later.

Thomas's work was important, but at this point the rediscovery of how to read Mayan long-count dates was not yet complete. It was one thing to be able to translate numbers from a Mayan to a modern notation, but in terms of the calendar, this information was almost useless until links could be found between their calendar and known events that had been recorded in our own calendar. This was achieved by one of the unsung heroes of Mayanology: Joseph Goodman. Goodman was another American entrepreneur, made rich through profitable investment in the Virginia City gold boom of the 1860s and '70s. Retiring to Fresno, California, in the 1880s, Goodman had time and money to pursue Mayan studies. Working with the latest material, supplied to him by his friend Alfred Maudsley – then the best field researcher in Mexico and Belize – he made three important discoveries. The first was that the Mayan calendar began with a start-date expressed in the 52-year cycle as 4 Ahau 8 Cumku.

This seemed on the face of it a rather surprising choice, since Cumku is the last but one month and its ninth day (as the numbers run from 0 to 19) is not even the end of that month. Of the day-names, Ahau, or Lord, seemed appropriate enough for the start of a new era but why begin with the number 4? These are questions that have perplexed Mayanologists ever since Goodman's discovery of the 4

Ahau 8 Cumku start-date. However, there is no disagreement that he was right.

Goodman's second discovery was that in place of the dot and bar notation for numbers, the Maya also on occasions used 'head variants', rather like our occasional use of the Roman numerals 'V', 'L', 'C' and 'M' to represent the numbers '5', '50', '100' and '1,000'. The Maya's heads are all different but each has quite recognisable characteristics. Some of them represented gods who were already associated with specific numbers, but why the others were chosen remains a mystery.

Goodman's third discovery was the most important of all. Working with Landa's *Relación* and other post-conquest documents, he established that day one of the long count, i.e. 0 baktuns, 0 katuns, 0 tuns, 0 uinals, 0 kins, 4 Ahau 8 Cumku, occurred in 3114 BC. This date was later refined by J. Eric Thompson to between 11 August 3114 BC and 13 August 3114 BC. There are arguments for or against each of the three dates, 11, 12 and 13 August, but few people disagree with what is now generally known as the Goodman–Martinez–Thompson correlation, or GMT for short. Because we have this correlation of calendars, Mayanologists are able to take any inscribed long-count date and make it intelligible in terms of our own calendar. As importantly, it also means that using suitable astronomical computer programs we can backtrack the skies to the specific day in question. As we will see later, this has led me to some interesting discoveries.

DECODING THE TEXTUAL GLYPHS

By 1920, the theoretical basis for translating numerical Mayan inscriptions into recognisable dates on the Gregorian calendar was fully in place. The work of collecting dates, begun nearly a century earlier by Stephens and Catherwood, was taken to its limits by Sylvanus Griswold Morley. Under the auspices of the Carnegie Institution, he spent decades exploring the forests of the Peten region of Guatemala searching out lost cities for their treasure trove of inscribed dates. These he would either photograph or carefully draw with a pencil, thereby building up an extensive collection of hitherto unknown material. It remained for J. Eric Thompson, Morley's academic heir and protégé, to rehabilitate the work of Goodman, propose the GMT correlation and establish a workable chronology for the Mayan civilisation. Unfortunately, Thompson, who embodied the British colonial approach towards native cultures, thought little of the Mayans' abilities when it came to the written word. In his book *The Rise and Fall of Maya Civilization*, he writes:

> I believe that the Maya had neither an alphabetic nor a syllabic writing except insofar as most Maya words are monosyllables. There is considerable use of a simple phonetic writing which might be described as an advanced form of rebus writing in that the picture has become so conventionalised that the original object is no longer recognisable . . .
>
> An example of old-fashioned rebus writing is supplied by the Mayan sign for 'count'. In Yucatec, the word *xoc* means 'to count', but it was also the name of a mythical fish which dwelt in the sky and to which worship was made. As the Maya had difficulty in rendering an abstract idea such as 'count' in glyphic form, they turned to rebus writing and used the head of the *xoc* fish as the glyph for *xoc* 'count'.[1]

This was the orthodox way of thinking in 1950, and it brought Mayanology no nearer to understanding the meaning of most non-numerical hieroglyphs than it had been in 1773, when Father Ordoñez visited the ruins of Palenque for the first time. To break the code, other minds were needed: minds uncluttered by the prejudice that now surrounded the whole subject of Mayan epigraphy. As it turned out, such minds were not to be found in the departments of Mayanology by now firmly established in Germany, the United States, France and Britain, but far to the east in the Soviet Union.

Yuri Knorosov was the last person anyone of Thompson's generation would have expected to become a leading expert on Mayan epigraphy. Born in 1922 in Kharkov, in what is now the Ukraine, the world in which he lived was many thousands of miles away physically from the jungles of Central America and light years intellectually. Yet it was his unexpected destiny – which Edgar Cayce, the sleeping prophet, would undoubtedly have linked to a past-life experience as a Mayan priest – to find the keys that would unlock the code of the non-numerical inscriptions.

The story began in the ruins of Berlin, at the end of the Second World War. Knorosov, then an ordinary soldier in the victorious Red Army, found himself at the German National Library, which was then on fire. Keen to find a trophy to take home with him, he saved one book from the flames. This turned out to be a single-volume edition of Villacorta's facsimiles of the *Dresden, Paris* and *Madrid* codices that was published in 1933. That might have been the end of the story except that, as luck or perhaps fate would have it, Knorosov was no ordinary conscript but a genius with a penchant for ancient languages. Following the war, he went back to Moscow University, where he researched Egyptology, Chinese, Arabic and the languages of India. At the time, he might not have guessed it but this grounding in comparative linguistics gave him the perfect base from which to undertake the translation of the Mayan texts. Perhaps, though, he did have an inkling of where his destiny lay and that was the reason why he rescued the Villacorta rendition of the Maya codices and not some other, on the face of it more valuable, book from the burning wreckage of the National Library.

As well as the Villacorta book, Knorosov also had available to him a copy of Bishop Landa's *Relación de las Cosas de Yucatán*. As a scholar keenly interested in ethnology, this would have been of great interest to him, for it contains details concerning the customs, dress, calendars and religious practices of the pre-conquest Maya of Yucatán. However, Landa's book, which was one of the gems rediscovered by Brasseur de Bourbourg, contains more than just ethnography: it was to turn out to be the Rosetta Stone of Mayanology.

Few people would disagree with the assertion that Landa's book is one of the most important documents we have in our possession for understanding the religious systems of the Maya. Indeed, it was this book which enabled the

Goodman–Martinez–Thompson correlation to be worked out, establishing the start of the Maya long-count calendar as *c.* 13 August 3114 BC, plus or minus one day.[2] More controversial was a short section Landa wrote on the Mayan 'alphabet', where he explains how they formed their letters and gives examples:

> These people also used certain characters or letters, with which they wrote in their books about the antiquities and their sciences; with these, and with figures, and certain signs in the figures, they understood their matters, made them known and taught them. We found a great number of books in these letters, and since they contained nothing but superstitions and falsehoods of the devil we burned them all, which they took most grievously, and which gave them great pain.
>
> Of their letters we give here an a, b, c, their cumbersomeness not permitting more, because for all their aspirations of the letters they use one character, and then for uniting the parts another, going on in this way *ad infinitum*, as in the following example. *Le* means a lasso, and to hunt with one; to write it with their letters, they wrote them with three, at the aspiration of the *l* the vowel *e*, put before it; in this they are not at fault, although they use the *e* if they wish to do so for definiteness. Example: *e l e l;* afterwards they put the syllable joined:
>
> *Há* means water; because the sound of the letter *aitch* is composed of a, h, before it, they put it at the beginning with *a*, and *a ha* at the end in this fashion:
>
> They also wrote in syllables, but in one and the other style: I only put it here in order to give a complete account of the matters of this people. *Ma in kati* means 'I do not wish', and they write it in syllables in this manner:

Here begins their a, b, c:

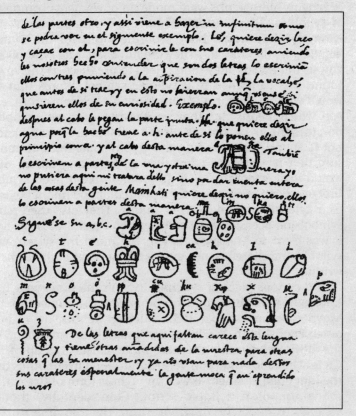

The letters that do not appear are wanting in this language; and they have others in addition to ours, for other things where they are needed. But they no longer use any of the characters, especially the young people who have learned ours.[3]

Read properly, Landa's words are unequivocal and provide the key to a correct translation of the Maya texts. Knorosov, who had an exceptionally wide understanding of linguistic systems, immediately recognised that Landa's alphabet was not consonants and vowels like our own Roman lettering or the Greek Cyrillic system used in Russia but, as Landa himself says, syllabic. Knorosov proposed that the majority of Mayan hieroglyphs were made up of elements composed of syllables. Thus, instead of having a single letter 'l', for example, they would have separate forms for the sounds 'la', 'le', 'li', 'lo' and 'lu'. They did have separate symbols for vowels but mostly they didn't need to use these as the vowel sound was contained in a syllabic symbol. Words that we would write in the form of consonant–vowel–consonant, e.g. *pop*, which means mat, they would write using two syllables: po-p(o). The 'o' part of the second syllable they would know not to sound when reading the glyph. Sometimes, to make sure that it was clear what they were talking about, they would use pictograms as well, for example a head symbol that everyone knew stood for a particular god might be used as a determinative next to his name. They also used rebus symbols on occasion, as Thompson had surmised. However, it was clear to Knorosov that in the main the system was syllabic and therefore not unlike many other early forms of writing.

In a curious piece of synchronicity, in the same year that Knorosov published his first paper on the subject of the Maya code, in October 1952, Michael Ventris, an equally gifted Englishman, made his first public announcement, on the BBC, concerning the enigmatic Linear-B script from Crete. The following year, Ventris published his findings that the language of Linear-B was an archaic form of Greek that made use of a syllabic script. Coincidentally, though visually completely different from the Maya script as revealed by Knorosov, the two hieroglyphic systems worked in almost exactly the same way. There were good reasons for this. Using a syllabary instead of an alphabet might seem cumbersome, and indeed it is, but it does have one advantage. Although the scribe needed to know many more signs and symbols than we do with our 26-letter alphabet, when it came to writing words he could get away with using fewer of them.

When Knorosov published the results of his research into the Mayan script, he at first met with opposition from orthodox scholars, especially J. Eric Thompson. In part, this was a political reaction towards a Soviet scholar who was assumed to be Communist in his sympathies. More to the point, Knorosov's work induced fear of the unknown. If it were shown that he was right in his analysis, it would turn Mayanology on its head and render redundant much of what Thompson thought he had achieved in over 30 years of study. This is, unfortunately, a common response of orthodox archaeologists towards new ideas, especially when these emanate from unqualified sources. It is, however, self-defeating. For in the end new ideas and theories, provided they are well founded, will always triumph over the old. That is the nature of progress and what gives dynamism to all the sciences, even when that science is archaeology and its field of research is ancient antiquities.

Though Thompson and others of his generation dismissed Knorosov's work as nonsense, other, younger minds were more open. Once his papers were translated into English and published in the West, his ideas began to attract a following. At first, such support was tentative, as archaeologists working in the field were all too aware of the influence wielded by Thompson and his circle and the possibly catastrophic effect on their chances of promotion if they overstepped the mark. Nevertheless, his ideas found some support and began to take root.

One champion who was yet to enter the ring was Tania Proskouriakoff. She was also Russian but, like Thompson, worked under the auspices of the Carnegie Institution. Seemingly out of loyalty to her colleague, she did not openly support her fellow Russian Knorosov until after Thompson had retired, but already, in the 1950s, she was making discoveries of her own. The most important of these stemmed from a close study of standing stelae at the Mayan city of Copan.

The current opinion at the time was that these represented priests. Proskouriakoff was able to prove that they were rulers and that each collection of such stelae gave the dates and represented different stages in the ruler's life: his birth, coming of age, accession and death. The importance of this discovery was that it confirmed what Stephens and Catherwood had proposed about Copan when they visited the site a century earlier. They had claimed that the inscriptions were not simply artistic ornamentation or vague pieces of religious iconography but that they recorded

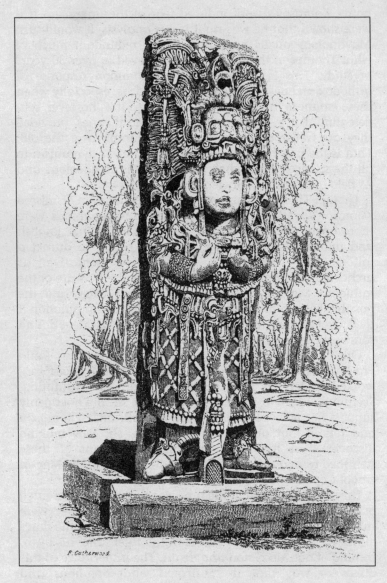

Figure 14: Front view of 'Stone Idol' from Copan

Figure 15: Rear view of same

the history and names of the people who built the city. Proskouriakoff had now proved that this was the case.

People began then to realise that by putting together the discoveries of Knorosov and Proskouriakoff it should be possible, once the exact language was identified, to read the monuments of the Maya and thereby recreate their lost history.

This was, however, but the beginning of what was to prove the most exciting time ever in Mayan archaeology. In 1952, the same year that Knorosov published his seminal paper on Mayan hieroglyphs, Alberto Ruz, an intrepid Mexican archaeologist, made a most extraordinary discovery inside the same Pyramid of Inscriptions that a century earlier had so inspired Stephens and Catherwood on their visit to Palenque. Looking at one of the flagstones that make up the floor of the temple, he realised that there was a strange anomaly. This particular stone was unlike all of the others in that it had a double row of holes cut into it, each plugged by a stopper. Suspecting that there might be something important concealed underneath the flagstone, he arranged for it to be lifted. To his amazement, he found what turned out to be a filled-in stairway.

It took four seasons of work to clear the staircase of rubble, thereby revealing a secret chamber at the bottom. This chamber contained grave-goods including the skeletons of six adults who were probably sacrificial victims. This in itself was a great discovery but there was much more to come. In the wall of the first chamber, which Ruz now realised was below the base-level of the pyramid, was what appeared to be a triangular door. After removing all the remaining archaeology from the first chamber, Ruz and his men set to work and prised open the door. Behind it they found a second, secret chamber. At 9 ft by 7 ft, this chamber was quite large. Around its walls were depicted, in stucco relief, nine figures, all in full ceremonial dress. Lying on the floor, exactly as they had been placed at the time the tomb was sealed, were a number of items of further archaeological interest. These included two figures carved out of jade and two beautifully moulded heads representing young men. However, these finds, important as they were, paled into insignificance when compared with the chamber's real treasure: a huge sarcophagus.

By far the most elaborate yet found anywhere in either of the Americas, it was found to contain a rich collection of funerary objects. These included jade beads, ear-spools, necklaces and rings. Most precious of all was an extraordinary mask – again made out of pieces of jade – which covered the face of the tomb's occupant. Though nothing remained of his flesh or clothing, it was clear from the grave-goods that he had been someone important: an *ahau*, or king. The skeleton of this man indicated that he was exceptionally tall – certainly in comparison with the modern Maya.

Even more significant was the lid of the sarcophagus. It measures 3.8 by 2.2 metres and weighs about 5 tons and is quite unlike any other yet discovered. Richly carved, most of its upper surface depicts what is clearly intended to be a mythological scene. At its centre is what appears to be a strangely dressed man at the foot of a symbolic tree. The man looks as though he is falling backwards, while the 'tree' has the appearance of a Roman cross, its cross-beams terminating in what looks like a pair of snakes' heads. Draped over the tree/cross is what looks like another serpent, this one with what appears to be dragons' heads at either end. Perched on top of the tree is a curious bird with a bifurcated tail.

This was the tantalising scenario in 1953, the year when Yuri Knorosov published his first paper on his method of decoding the Mayan script. It was an exciting time for Mayanology, and though Ruz didn't realise it at the time, Palenque was to continue to be the epicentre of many further discoveries. Ruz published his findings and almost immediately they caused a storm. From the state of the entombed man's teeth and bones, Ruz estimated that he was probably about 40 years old when he died. As for who he was, this was to remain a mystery for the time being. Meanwhile, on the basis of an inscription on the lid, Ruz referred to him as '8 Ahau'.

Since nature abhors a vacuum, this left the field clear for some fanciful theories to take root. The most famous of these was published in 1967 by a Swiss author, Erich von Däniken, in his first and most famous book, *Chariots of the Gods?*. Here he proposed that the Earth was visited by extraterrestrials

Figure 16: Drawing of the Lid of Palenque

and that these were the so-called 'gods' of the ancient world. He pointed to the Lid of Palenque as evidence of such a visitation; it represented, he claimed, a spaceman at the controls of his craft. Although *Chariots of the Gods?* was greeted with scepticism by the scientific community, it was extremely popular with a public by now used to the idea of men in space and eagerly anticipating the first lunar landing. Unlike the dry academic journals in which archaeologists presented their findings, it reached a vast audience worldwide. Until then, most of these people had never heard of the Lid of Palenque and had little or no interest in Mayan archaeology. Whatever may be said about von Däniken – and over the years some pretty bad things have been said – there is no disputing that *Chariots of the Gods?* changed all that. Suddenly, the ruined cities of the Maya, Palenque in particular, became popular places to visit. Not content to take scientific rebuttals of his ideas at face value, people wanted to walk in von Däniken's footsteps and make up their own minds about his spacemen gods.

In 1973, nearly 20 years after the publication of Knorosov's seminal paper, a major conference took place at Palenque that was to change the course of Mayanology forever. Called *Mesa Redonda de Palenque* – 'Round Table of Palenque' – it brought together an amazing group of people: artists, epigraphers and archaeologists, young and old, amateur and professional. What they all had in common was enthusiasm and a desire to break through the barriers holding back research into the Maya. In particular, they wanted to establish whether Knorosov's methodology really worked and, if so, whether it could be used to decode the many texts that covered such famous buildings as the Pyramid of Inscriptions. Though coming from different fields and with quite diverse perspectives, it is as though the people gathered together, many meeting for the first time, were all souls sent to Earth for the express purpose of revealing the secret history of the Maya. Such was the impact of this conference and the other *Mesa Redonda* gatherings that were to follow that it is tempting to think that the delegates were reincarnated Maya who had lived earlier lives as Palenque scribes. Who knows, perhaps they had worked together before sculpting the very texts that they were now engaged in decoding? It would have been interesting to hear what Edgar Cayce would have had to say about this.

Among the attendees was a young artist called Linda

Schele. A junior teacher at the University of Mobile in the southern state of Alabama, she had first visited Palenque in 1970 and fallen in love with it on first sight. Also attending was Peter Matthews, an Australian undergraduate from the University of Calgary in Canada. Peter's tutor there was David Kelley, a remarkable scholar who, to the annoyance of J. Eric Thompson, was a convinced diffusionist and had been quietly building upon the foundations laid down by Knorosov. Well instructed by Kelley, who couldn't himself attend, Matthews brought with him an encyclopaedic knowledge of the hieroglyphs and a notebook containing every known hieroglyphic date inscribed on the monuments of Palenque.

The meeting of Schele and Matthews turned out to be possibly the most fortuitous in the history of Mayan research. Working together, they set out to assemble a dynastic history of Palenque, a task only now becoming possible. The result of their efforts was the first family tree for a dynasty of Mayan rulers. Using Knorosov's methodology as developed and amended by other scholars, they ascertained that the name of the man buried in the tomb was not '8 Ahau', as Alberto Ruz believed, but *Pacal*, a name meaning shield. His title was *Makina*, a word meaning great sun lord. Further research revealed that this king had a son called *Chan-Bahlum*, or snake-jaguar, and a mother called Lady *Zac-Kuk*, or white quetzal. She was the daughter of an earlier Pacal, whom they now began to refer to as Pacal I to distinguish him from his grandson Pacal II, or Pacal the Great.

It was an astounding piece of detective work, though unfortunately it put them at odds with the archaeological establishment as represented by Alberto Ruz. He was now the head of Mexico's I.N.A.H., the department responsible both for the preservation of ancient monuments and for the issuing of licences to carry out digs. He was adamant that the man in the tomb, whom he still called 8 Ahau, was only around 40 years old when buried. Accordingly, he branded the new decipherers 'fantasists' and did his best to undermine the credibility of their work.

J. Eric Thompson died in 1975 and Alberto Ruz in 1979. With their departure from the scene, there was no one of any stature left to oppose the tidal wave of new thinking that was changing the face of Mayan studies. Within a few years, linguistics had virtually taken the place of dirt archaeology as the means of finding out about the Maya. The ability to read and translate hieroglyphic texts was now more important than almost anything else. In 1990, Linda Schele and archaeologist David Freidel, whose knowledge and skills well complemented her own, co-authored a book entitled *A Forest of Kings*. It was to be a seminal work, for though her ideas were well known within the narrow circle of Mayan specialists, this was the first presentation of them in popular form for the general public. I had it in mind when, continuing my journey, I returned to Palenque myself.

CHAPTER 6

A New Vision of the Maya Creation

The coach wound its way along a narrow road that threaded through fields of pasture, the property of local *rancheros*. The scene passing by my window seemed curiously out of place. In places, scattered here and there, were trees. They looked a lot like the deciduous trees of home: silver birches, ashes, beeches and oaks. They were not at all what I was expecting to see on the fringes of a jungle. The fields, too, green from tropical rainfall, looked more like Kent than Chiapas. I felt like I was in a dream and that at any moment I would wake up to find myself back home, parked outside an ancient pub or pretty tea shop.

Presently, with an abruptness that was startling, ranch-land gave way to primal forest and we found ourselves in relative darkness beneath a canopy higher than any cathedral nave. The coach parked by the gate to what was evidently a park, though as this was on the crest of a small hill, we could not yet see what lay beyond. Along the side of the road on the approach to the gate were stalls selling souvenirs: everything from T-shirts and dresses to model pyramids and posters. These, however, did not catch my attention, which was drawn instead to a small group of men clustered right next to the gate itself. Dressed differently from the stall-holders, they wore long white smocks that reached down nearly to their feet. Their hair was black and long. It hung loosely down their backs, further adding to the impression that they had just got out of bed. Small in stature, red of skin and lean to the point of malnutrition, it suddenly dawned on me that these must be the Lacandon Indians I had read about in books but till then had never seen for myself.

The Lacandons, like the Indians of the Amazon rainforest, still live in the jungles of Chiapas. Descendants of the ancient Maya whose pyramid cities so fascinate us today, they prefer subsistence-living in the jungle to absorption. Their dogged determination to remain separate from the mainstream, to cling onto the 'old ways' of their ancestors, makes them of great interest to tourists as well as anthropologists. Yet, in one of those curious ironies of life, it is this benign curiosity that threatens their independence as never before. The few dollars that they earn selling knick-knacks to tourists are mostly spent buying goods in the local towns. Consequently, they are gradually being drawn into the global economy. The older folk may not like it but who could blame them if the young Lacandons, having seen what the towns have to offer, forsake the jungle for more regular employment? This is a dilemma facing traditional communities everywhere in the Americas but nowhere is it more acute than here in what remains of the great forests of Chiapas.

Leaving the Lacandons to themselves, I walked on through the gate. Presently, I found myself standing on a rough-grass 'lawn' under the shade of a young ceiba tree. I felt overjoyed to be back in Palenque, this time better equipped to understand and appreciate the marvel that lay before me. In front of the ceiba was Palenque's most famous building: the Pyramid Temple of Inscriptions. As it was this building more than any other that I had returned to see, I felt a frisson of excitement. If anything, it seemed taller and even more impressive than it had on my last visit. Whether this was because I was in less of a hurry this time or because I knew rather more about it was hard to say. Either way, it felt great to be back, to savour once more the sense of walking in the footsteps of del Rio, John Stephens, Frederick Catherwood, Count Jean Frederick Waldeck, Alfred

Maudsley, Sylvanus Morley and all the other great pioneers of Mayanology.

Because the Pyramid of Inscriptions is so pleasing to look at, I spent some minutes standing in the shade of the ceiba tree in quiet contemplation. The reason for its appeal is not hard to discern and is revealed in the figures given by Stephens for the dimensions of the temple: 76 ft wide by 25 ft deep. In other words, it is three times as wide as it is deep. The proportion of three to one is naturally harmonious and although the pyramid is stepped, its facade is naturally triangular in shape; the average slope of the pyramid uniform at about 60°. At the time it was built, the roof of the temple on top of the pyramid would have supported a 'comb'. This would have been an open-work stone trellis. The primary purpose of the roof-comb was to supply vertical weight and thereby stabilise the roof. Although this roof-comb is no longer present, as there were similar examples to be seen on some of the other buildings at Palenque I could imagine what it probably looked like. The apex of the triangle defining the slope of the sides almost certainly dictated the height of the top of the now lost roof-comb. This invisible but implied triangle is the principal reason, I believe, why the building looks so attractive to the human eye. That the ancient Maya knew about harmonic proportion and used it to their advantage is silent testimony to their intelligence and the advanced nature of their civilisation.

Leaving the shade of the ceiba, I walked over to the pyramid and then up the narrow-stepped staircase that runs from ground level to the central entrance of the temple. At 65 ft in height, it was not an easy climb, especially in the hot and humid conditions of Palenque. By the time I reached the top, I was out of breath and my shirt was clinging to my back with perspiration. I paused for a few minutes to admire the view and examine the tables of inscriptions that had so impressed Stephens and Catherwood in the 1840s. However, seeing these was not my only reason for climbing the pyramid: I wanted to revisit the tomb of Pacal.

Following in Alberto Ruz's footsteps, I began the perilous descent down the slippery stairway. Dank and with its walls covered in lichens, it was even more humid than I remembered from my last visit some two years prior to this one. Having been there before, I knew what to expect when I reached the bottom, and indeed nothing seemed to have changed. Behind a locked iron-grille doorway could be glimpsed the tomb and in the centre of it the great sarcophagus. On its sides were carved other richly adorned figures, the ones Linda Schele had identified as being Sun Lord Pacal's ancestors. It was covered by what is the tomb's most enigmatic and, archaeologically speaking, most important find: the famous lid. As I had been here before, I knew what to expect. However, this time, having read *A Forest of Kings*, I knew a great deal more about it.

That the Lid of Palenque was intended to illustrate Mayan religious ideas concerning death and the afterlife is clear from its context. However, without the tools to decipher it, its subtler message was beyond the understanding of its discoverers. This left the field clear for others to fill the gap with their own interpretations, the most famous being that furnished by the extraterrestrialist Erich von Däniken in *Chariots of the Gods?*. As we have seen, according to him the figure shown on the lid was an ancient astronaut, at the controls of an exotic spacecraft. More cautious voices pointed out that since the Maya were basically a Stone Age people who did not even have wheeled vehicles, still less the sort of advanced technology needed to make space travel even a possibility, this was unlikely. However, as implausible as von Däniken's explanation was, it captured the imagination of the 'space-age' generation and his book *Chariots of the Gods?* became one of the most remarkable bestsellers of all time.

Maurice Cotterell, my co-author on *The Mayan Prophecies*, had his own theories concerning the Lid of Palenque. He argued that the lid contained a secret code – or rather a number of secret codes – that could only be appreciated by overlaying outline sketches of the lid printed on transparent acetates. Shifting these acetates around, one over the other, revealed pictures that he believed represented secret teachings of the Maya. According to him, the human figure at the centre of the lid was not a man at all but rather a representation of the Mayan version of Chalchiuhtlique, the goddess of standing water. The bird sitting on top of the tree was, he claimed, the Mayan equivalent of the Aztec god Ehecatl, who symbolised the wind as an aspect of the 'Plumed Serpent' Quetzalcoatl. At the foot of the tree were figures symbolising two other Aztec gods: Tlaloc (*Chaac*), the rain god, and Tonatiuh, the sun god. He saw the whole lid as the Mayan equivalent of the Aztec calendar stone. At the time we wrote *The Mayan Prophecies*, I thought this was a plausible interpretation. Now that I know rather more about Mayanology, I no longer believe this to be the case.[1] The truth is, in fact, far more exciting.

The first *Mesa Redonda* conference of 1973 was followed up by many others held either at Palenque or in the United States. Using Knorosov's technique of breaking the hieroglyphs down into syllables, the scholars were able, bit by bit, to decipher the complex texts recorded on the Pyramid of Inscriptions and other Palenque buildings. The result of this extraordinary collaboration was a complete revolution in thought. It became clear that Pacal, or Lord Sun Shield, was born on 26 March 603, came to the throne on 22 March 615 and died on 31 August 683. The inscriptions on the sides of his tomb give further details concerning his ancestors going back for eight generations. The actual tomb of Pacal, planned by himself but finished off by his son, contains many references to his family tree. Painted stucco figures on the walls and around the tomb itself, once thought to have been representations of the 'Nine Lords of the Night', were actually Pacal's ancestors.

This is the context in which we now have to study the imagery of the Lid of Palenque. The figure at the centre is neither one of von Däniken's spacemen, nor Cotterell's goddess of water. It is clear from the inscriptions that it in fact represents Lord Pacal himself, who, according to Schele, is pictured in the act of falling into the jaws of death. A glyph on the lid says *och beh*: 'he entered the road'. This, she said, was a reference to the Milky Way, which the Maya called *sak beh*, or the white road. The reasoning behind this is the presence of the cross-shaped tree. This tree, with its flowers, is intended to symbolise a giant ceiba tree, which for the Maya represented the Tree of Life. Along its trunk are mirrors, which the Maya even today hang from the ceiba's

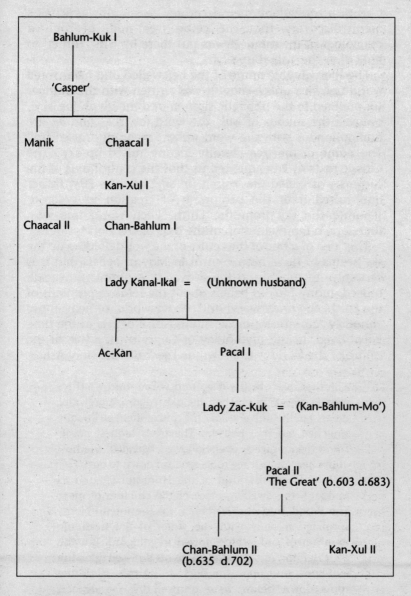

Figure 17: The family tree of Pacal (after Linda Schele's A Forest of Kings*)*

Christian equivalent: their decorated crosses. It was Schele's great discovery that the ceiba tree had, at root, a cosmological meaning: it was put there by 'First Father' to hold aloft the 'raised up sky'.

The hieroglyphic name of the bejewelled and bemirrored world tree was *wakah-chan*. It was written with the number six prefixed to the phonetic sign *ah* and the glyph for 'sky', because the sounds of *wak*, the word for 'six', and *ah* are homophonous with the word *wakah*, meaning 'raised up'. The name of the tree literally meant 'raised up sky'. The Classic texts at Palenque tell us that the central axis of the cosmos was called the 'raised up sky' because First Father had raised it at the beginning of creation in order to separate the sky from the Earth. Each world tree was, therefore, a representation of the axis of creation.[2]

This idea of a tree at the centre of the world holding up the sky seems to have been central to Mayan beliefs, but it is certainly not unique to the Maya: it is almost universal. Indeed, many Mayan beliefs about the relative positions of the Earth, the underworld and 'above-world' of heaven are curiously reminiscent of Germano-Norse beliefs at the time Pacal lived. In the mythology of Germany, the tree at the centre of the world was known as *Yggdrasil*, the Holy Ash.

> Allfather dwelt in the deep and willed, and what he willed came to pass. Then the ash Yggdrasil grew up, the tree of the universe, of time and of life. The boughs stretched out into heaven. The tree's highest point, *Lärad* (peace-giver), overshadowed *Walhalla*, the hall of the heroes. Its three roots reached down to dark *Hel*, to Jotunheim, the land of the Hrimthurses, and to Midgard, the dwelling-place of the children of men. The World Tree was evergreen, for the fateful Norns sprinkled it daily with the water of life from the fountain of Urd, which flowed in Midgard. But the goat Heidrun and the stag Eikthynir browsed upon the leaf-buds, and upon the bark of the tree, while the roots down below were gnawed by the dragon Nidhögg and innumerable worms: still the ash could not wither until the last battle should be fought, where life, time and the world were all to pass away.

Figure 18: Yggdrasil, the Holy Ash

> So the eagle sang its song of Creation and Destruction
> on the highest branch of the tree.[3]

In pre-Roman Britain, the sacred tree was the oak, as indeed
it is in the ancient *Edda* of Finland:

> Says one of the Finnish runes: 'Long oak, broad oak.
> What is the wood of its root? Gold is the wood of its
> root. The sky is the wood of the oak's summit. An

enclosure within the sky. A wether [castrated ram] in
the enclosure. A granary on the horn of the wether.'[4]

There is some evidence that in Mesopotamia too, at the time
when the *Epic of Gilgamesh* was probably composed (some
time during the third millennium BC), the Tree of Life was
symbolised by an oak. In later periods, it was more often
depicted as a date-palm. In the Bible, of course, the Tree of
Life stood in the middle of the Garden of Eden. It may or
may not be the same as the 'tree of knowledge of good and
evil', whose tasty fruits proved irresistible to Adam and Eve.

That the ceiba tree was of similar importance to the Maya
as the ash was to the Germans or the oak to the Druids is not
in doubt. The cosmological importance of the ceiba was
remarked upon by J. Eric Thompson himself: 'Of the ceiba
tree, still considered sacred, many legends and superstitions
survive, although the old cosmological beliefs have largely
disappeared under the impact of Christianity.'[5] One reason
for the ceiba's eminence, besides its enormous size when
fully mature, was its economic importance. The wood was
used for making canoes, and its fruits, which are filled with
cotton-like fibres, are still used for stuffing mattresses. Linda
Schele's genius was that she was able to see beyond such
practical, utilitarian properties of the tree to the
astronomical facts underlying the mythology. The first clue
to what this might be came not from the inscriptions written
on the walls of Palenque but from an old pot. Indeed,
Mayan ceramics have turned out to be a key source of
information concerning the mythology of the ancient Maya.

In 1971, an exhibition of Mayan pots, the first of its kind,
opened in the Grolier Club New York. The exhibition was
primarily intended to display Mayan texts. However, since
most of these are inscribed on blocks of stone weighing
many tons, there were few of these to be found in New York.
The inclusion of ceramics was the suggestion of Michael D.
Coe. He was a pivotal figure bridging the gap between old
and new thinking. As an archaeologist, he had worked with
both Thompson and Morley but he was, nevertheless, open-
minded enough to embrace the tide of new thinking
represented by the work of Yuri Knorosov and was a firm
friend of the Round Table scholars. Coe put together an

amazing selection of Classic Period pots, some of doubtful ownership, which were loaned for the occasion by private collectors and dealers in antiquities. Lending these pots was conditional on a promise that at the end of the exhibition they would be returned to their current owners with no questions asked about how they were obtained. Since many of the pots were thereafter likely to be withdrawn from public view, it meant the exhibition was likely to be the only time most people would ever see them. Instead of raising a stink about the private trade in antiquities, Coe realised that it was better to treat this as a golden opportunity to record such information as they carried in the form of hieroglyphic inscriptions and pictorial representations. Accordingly, he arranged for graphic 'rollouts' of the pots to be made by an artist. These drawings were included in a catalogue of the exhibition that was published in 1973 under the title *The Maya Scribe and his World*.[6] Coe published more books on Mayan pots, including *Lords of the Underworld: Masterpieces of Classic Maya Ceramics* (1978) and *The Hero Twins: Myth and Image* (1989). In 1980, his illustrator, Justin Kerr, published a book of his own – the first of six volumes.[7] Additionally, in 1981, Francis Robicsek and Donald Hales published a further work entitled *The Maya Book of the Dead: The Ceramic Codex*.

Thus it was that in November 1991, when Linda Schele and her co-author David Freidel attended a conference in Chicago, there was no shortage of books containing photographs of Mayan ceramics. During the conference, Freidel got into conversation with an epigrapher called Bruce Love. They discussed the *Paris Codex* and how this contains pictures of zodiacal constellations, including one that looks like a scorpion.[8] Freidel mentioned this to Schele, telling her that he remembered seeing a picture of a scorpion on a Mayan pot that featured a man with a blow-pipe shooting a bird out of a tree.[9] Could this scorpion, he suggested, represent the same constellation we know as Scorpio?

Schele at first resisted this idea but came back to it later when she remembered seeing the constellation of Scorpio positioned over the collection of Palenque temples known as the cross-group that lie to the south of the Pyramid of

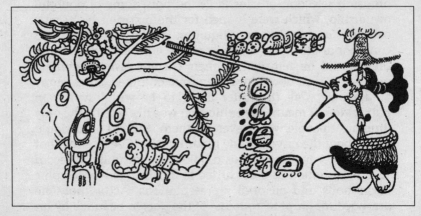

Figure 19: Hunahpu shooting a bird, with a scorpion by the tree

Inscriptions. She examined a star-atlas to see where the Scorpio constellation stands relative to other important star-groups. It was then that she made a surprising discovery: as it rose, so too did the Milky Way, which then straddled the sky from south to north. Could it be, she wondered, that the wakah-chan of the Maya, the 'raised up sky' tree, represented the Milky Way? If this were so, then the scene depicted on the pot of a scorpion placed at the base of the tree was a cosmic diagram. It symbolised the stars of Scorpio in relation to the Milky Way. Since the Milky Way was called sak beh, or white road, by the Maya, this would also explain the curious hieroglyphic inscription och beh, 'he entered the way', that was written on the side of the Lid of Palenque. Schele reasoned that Pacal's descent into the open jaws of death was meant to represent the beginning of his journey to the stars of the Milky Way.

Continuing the cosmological analogy, she proposed if the tree itself symbolised the Milky Way, then its branches, which resemble the cross-beam of a crucifix, were meant to be the celestial equator. Twined around the top of this 'cross-tree', in a way reminiscent of the Bible story of Moses raising a cross and serpent in the desert, was a double-headed serpent. This, she suggested, represented the ecliptic: the pathway followed by the Sun on its annual course through the sky. Again this was reminiscent of European mythology,

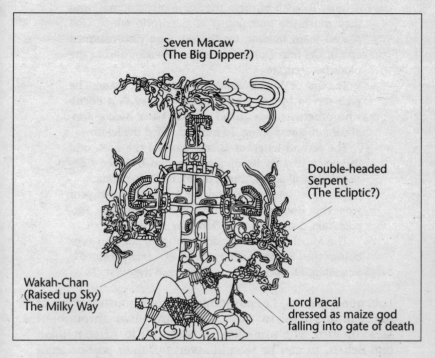

Figure 20: The Mayan tree of life from the Lid of Palenque, as defined by Linda Schele

where the ecliptic was also sometimes symbolised by a serpent biting its own tail.

Schele now turned her attention to the way the Maya delineated space. In Bishop Landa's account concerning the beliefs of the Yucatecan Maya, we learn that they paid careful attention to the four cardinal directions. Each direction was signified by a particular colour and associated with its own god.

> Among the multitude of gods worshipped by these people were four whom they called by the name *Bacab*. These were, they say, four brothers placed by God when he created the world, at its four corners to sustain the heavens lest they fall. They also say these Bacabs escaped when the world was destroyed by the

deluge. To each of these they give other names, and they mark the four points of the world where God placed them holding up the sky, and also assigned one of the four Dominical letters to each, and to the place he occupies . . .

The first of these Dominical letters, then, is *Kan*. The year served by this letter had as augury that Bacab who otherwise was called *Hobnil, Kanal-Bacab, Kan-pauahtun, Kan-xibchac*. To him belonged the south.

The second letter, of *Muluc*, marked the East, and this year had as its augury the Bacab called *Can-sicnal, Chacal-bacab, Chac-pauahtun, Chac-xibchac*.

The third letter is *Ix*, and the augury for this year was the Bacab called *Sac-sini, Sacal-bacab, Sac-pauahtun, Sac-xibchac*, marking the North.

The fourth letter is *Cauac*, its augury for that year being the Bacab called *Hosan-ek, Ekel-bacab, Ek-pauahtun, Ek-xibchac*; this one marked the West.[10]

This quadripartite, i.e. partitioned in four quarters, view of the world seems to have been universal throughout Mesoamerica. The same idea was central to Aztec religion and beliefs, as can be seen in several of their post-colonial books, notably the *Codex Borgia*. In the Mayan tradition, each of the cardinal directions was associated with a particular colour. East was associated with the colour red (*chac*), north with white (*sak*), south with yellow (*kan*) and west with black (*ek*). According to Eric Thompson, there could be four 'trees' associated with the cardinal directions and maybe a fifth placed at the centre:

At each of the four sides of the world (or perhaps at each side of one of the heavens) stood a sacred ceiba (the wild cotton tree), known as Imix ceiba, and these trees, too, were associated with the world colours. They appear to have been the trees of abundance, from which food for mankind first came; their counterparts in Aztec mythology helped to sustain the heavens. In a ritual of the four world directions in the *Book of Chilam Balam of Chumayel*, we read:

The red flint is the stone of the red Muzencab [the sky-bearer who also functioned as a bee god]. The red ceiba of the dragon monster is his arbour, which is set in the east. The red bullet tree is their tree. The red sapodilla, the red vine . . . Reddish are their yellow turkeys. Red toasted corn is their maize.

The rotation of the directions follows this quotation, each with associated deities, flora and fauna of the required colour. On each tree perched a bird of the requisite colour. There is reason to believe that a fifth green tree was set in the centre.[11]

By now, convinced that cosmology was the key to understanding the strange symbols depicted on the lid, Linda Schele turned her attention to other graphic elements. The most important of these was the bird which is shown standing on top of the tree. As I pointed out earlier, in *The Mayan Prophecies* Maurice Cotterell identified this as the Mayan equivalent of Ehecatl, the Aztec god of wind and a personification of the god Quetzalcoatl. Schele was of a different opinion. She identified him as *Itzam-yeh*, a bird god who appears in many contexts and symbolised the magical power of the Maya high god *Itzamna*. The word *itz* means magical stuff, the Mayan equivalent of the Chinese word *chi*, the Indian word *prana* or the Greek word *pneuma*. Itzam means shaman or sorcerer, i.e. someone who works with itz. Itzamna, one of the high gods of the Maya, was the archetypal or great sorcerer. According to Schele, the bird Itzam-yeh represented the *way*, or companion spirit, of *Itzamna*, rather as in Christianity the dove symbolises the Holy Ghost, or *pneuma hagion*, the divine force emanating from God that gave Jesus his powers of healing. Thus, the bird at the top of the tree on Pacal's lid could therefore be said to represent the presiding shamanic power of the god Itzamna.

This, however, was not her only thought on the subject. Schele was intrigued by the obvious connection made by the modern-day Maya between their conception of the universe and the structure of their houses. These are nearly always

Hearth stones

Figure 21: A typical Mayan house (after R. Wauchope, Modern Maya Houses*)*

oval in shape and orientated towards the four cardinal points.

Inside the house there is always an open fireplace on which the women of the house cook. This is the heart of any home and consists of three large stones placed to form an open triangle. Schele discovered that these three stones were thought of as having a cosmic equivalent. A stele from Quirigua, a Classic Mayan city in the eastern fringe of what is now Guatemala, told how creation began with the setting of these three stones by the gods. One stone – called the 'Jaguar-throne-stone' – was placed by two elderly gods known to Mayanists as the 'Jaguar Paddler' and 'Stingray Paddler'. A second, the 'Snake-throne-stone', was set by a god called *Ek-Na-Chak*. The third, the 'Waterlily-throne-stone', was set in place by Itzamna himself. Consulting Dennis Tedlock's translation of the *Popol Vuh*, she found a further reference to this stone-setting. According to Tedlock's sources, the three hearth-stones equated with the three southernmost stars in Orion: Alnitak, Saiph and Rigel. These form a large triangle in the sky similar in shape to the Mayan hearth at the centre of every house. The apex of this triangle is the star Alnitak, which is also the most southerly of the three stars making up Orion's Belt. The Maya evidently likened these belt-stars to a fire-drill (the early way of starting a fire by causing friction between a stick and another piece of wood). Below them, at the centre of their cosmic hearth, is a grouping of stars which today we call Orion's Sword. This star-group contains M42: the 'Horse-Head' Nebula. Seen through a powerful telescope, this dust-

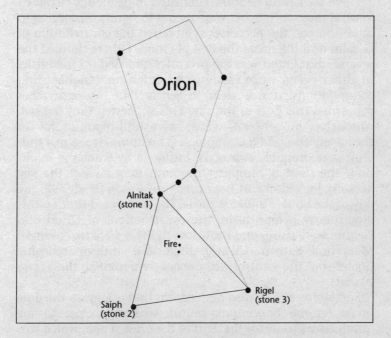

Figure 22: Drawing of Orion with stars of the hearth highlighted

cloud is one of the most beautiful objects in the sky. We now know that this area of the sky is a nursery where new stars are being born, but to the Maya it was the smoky fire at the centre of a cosmic hearth.

The creation myth of the Maya and its connection with the Orion constellation took a further turn when another of Schele's students asked her about a possible connection with a mural in one of the temples at the Classic Maya site of Bonampak. This set her off on a fresh and unexpected line of research. The mural in question, composed of possibly the most beautiful Mayan paintings to have survived to our time, features a 'judgement scene' commemorating a battle that took place on 6 August AD 792. Above the mural itself is a row of enigmatic cartouches. The first of these represents two peccaries, or Mexican pigs, in the act of copulating. The last of them shows a turtle with three stars on its back.

There seemed to be some confusion in academic circles as to what these cartouches were meant to signify. According to some sources, the peccaries symbolised the constellation of Gemini and the turtle the Belt of Orion. Others claimed the reverse: that Orion was the peccaries and Gemini the turtle. In either event, since Gemini and Orion are neighbouring constellations, it was clear to Schele that the cartouches concerned this part of the sky. This suggested that the two cartouches in between, which were anthropomorphic in character, also had a cosmological meaning. It seemed that they were probably star gods. Intrigued by what she might find, she used a computer program to see what the sky actually looked like at Bonampak on 6 August AD 792: the day of the battle. To her surprise, she discovered that on this day, shortly before dawn, the constellations of Orion and Gemini were rising and that close to them were two planets: Mars and Saturn. Clearly, then, the anthropomorphic figures on the central cartouches represented these two planets.

Something Schele also noticed was that 6 August, the date given on the Bonampak mural, was very close to the 'anniversary' date for the start of the present age, which was said to have begun on 13 August 3114 BC. This date, she knew, is recorded on Stela C from the Mayan city of Quirigua. This stela says that the act of laying the three stones took place at the completion of 13 baktuns, on a day called 4 Ahau 8 Cumku. This is not the only place where we meet this date, which seems to have been universally recognised throughout the Mayan world.

At Palenque, the most important record concerning the dates of creation is to be found in the Temple of the Cross. This is one of a small group of temples built by Pacal's son Chan-Bahlum and was where I went after leaving the Temple of Inscriptions. Arriving at the foot of the pyramid

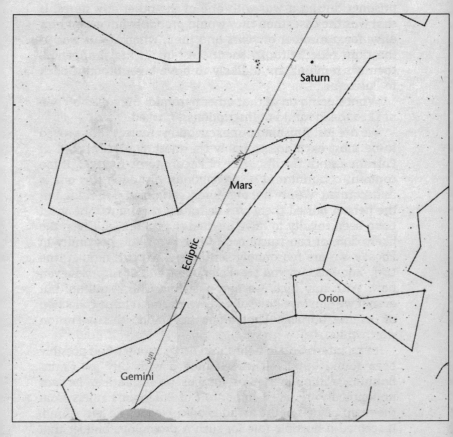

Figure 23: The sky as it looked at dawn at Bonampak on 6 August AD 792

on which this temple stands, I was relieved to discover that, although an archaeological excavation was in progress, the main structure was still open to visitors. The lady in charge of the dig was very friendly and I asked her what they were hoping to find. 'Oh,' she said, 'it's a long shot but we hope we might discover the tomb of Chan-Bahlum himself. After all, he had to be buried somewhere and this is his most important building. It celebrates his accession as king. Of course, he might not have been buried in Palenque at all. There is some evidence that he was taken prisoner during a war with one of the other city states. If that was the case, then they would probably have kept him alive for a number of years and then, when Venus was in the right place, ritually sacrificed him. If that happened, then his remains are unlikely to have been brought back to Palenque.'

'What's going on at that other pyramid: the roped-off one next to the Pyramid of Inscriptions?' I asked.

'We are just finishing our excavations there. It has proved to be most exciting, probably the most significant find at Palenque since the discovery of Pacal's tomb in 1952. It too contains a burial and a sarcophagus, although this one is undecorated. We are not absolutely certain who she was, but the person buried there was definitely a woman. She must have been royalty to have merited such a burial. Given the likely date of construction and her pyramid's proximity to Pacal's, we are reasonably certain she was his mother: the Lady Zac-Kuk. So, you see, Palenque is full of surprises even now. With any luck, we may just find Chan-Bahlum, but even if we don't we have already recovered a large collection of incense burners. That alone has made this operation worthwhile.'

It was interesting to hear that Pacal's mother had possibly been found but I felt myself moved with pity for Chan-Bahlum. Although not as famous as his father, he was undoubtedly the most important of Palenque's rulers from the point of view of the archaeological legacy he left behind. It seemed a horrible fate for such a prodigious builder that he should end up being sacrificed in a distant land. Whether or not this was his fate, which in the Classic Period was by no means unusual for a Mayan king, we have much to thank him for. More than anyone, it is he who was responsible for turning Palenque into the archaeological treasure trove that it is today. For in addition to finishing off the Pyramid of Inscriptions and ensuring his father's body was properly buried, he undertook a remarkable building programme of his own. The most impressive of his buildings was the Cross Group of Temples which, because they are richly adorned with both texts and pictures, provide a unique insight into Mayan life. It is these buildings more than anything else that have confirmed the connection between Mayan religion of the Classic Period and what we know from later texts such as the *Dresden Codex* and *Popol Vuh*.

Leaving the archaeologist to her work, I walked on up the stairs to the crown of the little hill on which stands the Temple of the Cross itself. It was much as I remembered it from my previous visit, except that it seemed somewhat larger. The temple provided welcome shelter from the sun, which was by now extremely hot. I paused for a moment to examine the inside of the corbelled roof. It seemed strange to think that it had survived for over 1,300 years, which means that it is older than almost any church in Britain. Leading from this outer room was a doorway to the inner sanctuary, the adytum of the building. On either side of this doorway were near life-size reliefs. The left-hand one showed King Chan-Bahlum himself, dressed in all his finery. On the other was the unmistakable figure of the 'Smoking God', who Mayanists often refer to as God L.

God L is associated with the underworld kingdom of *Xibalba* and also in some way presided over the day of creation of the present age: 4 Ahau 8 Cumku. Both Chan-Bahlum and 'the smoker' were facing inwards, so that I had to pass through their line of sight to gain entrance to the inner sanctum. The wall facing me was dominated by a much larger pictorial relief flanked on either side by panels carrying long texts in hieroglyphics. Depicted on the large relief was a scene reminiscent of the Lid of Palenque. It showed King Chan-Bahlum standing to the right of a tree that looked almost exactly the same as the one on the Lid. On the left-hand side of the tree was a much smaller figure. He was dressed in all the finery of a king and, according to Schele, was identifiable as Chan-Bahlum's father Pacal.

Leaving the Temple of the Cross, I walked over to the nearby Temple of the Floriated Cross. Very similar in design to the other, though not in as good condition, it too contained an inner sanctum with a large relief picture and accompanying text. Again, at the centre of the plaque was a 'tree' with the same sort of bird sitting on top. However, this time the tree was depicted as a maize plant rather than a ceiba. The corn-cob fruits of this tree were human

Figure 24: Smoking God from Temple of the Cross, Palenque

LEFT: Plate 1. Statue of Coatlique in Mexico Museum

BELOW: Plate 2. Teotihuacan: Way of the Dead alignment seen from Pyramid of the Moon

ABOVE: Plate 3. Teotihuacan: author on Way of the Dead with Pyramid of the Moon framed by Cerro Gordo

Plate 4. Teotihuacan: Pyramid of the Sun seen from below

Plate 5. Palenque: Pyramid of Inscriptions

Plate 6. Palenque: palace with viewing tower

Plate 7. Palenque: author on Pyramid of Inscriptions with Pyramid of the Cross in background

Plate 8. Palenque: Pyramid of the Foliated Cross

Plate 9. 'Aztec' flyers

Plate 10. Comalcalco: 'power station' pyramid

Plate 11. Comalcalco: sample brickwork

Plate 12. Comalcalco: 'Afghan warlord' head

Plate 13. Comalcalco: crossbones tile

Plate 14. The author with J.J. Hurtak

Plate 15. Palenque: Canamayte-style
plinth for shadow-pillar

Plate 16. Uxmal: open-maw temple
on top of Pyramid of the Magician

Plate 17. Chichen Itza: Pyramid of Kukulcan

Plate 18. Equinox serpent appearing
on Pyramid of Kukulcan in 1998

Plate 19. Dzibilchaltun: canamayte platform
and sun-stone on the east–west Sak Be

Plate 20. Carlos Barrios
performing a healing

Plate 21. Author (right) with Mayan Elder Carlos Barrios

Plate 22. Oaxaca: half-alien sculpture

Figure 25: Tablet on the back wall of the Temple of the Cross

heads, a reference to the Maya myth that the gods fashioned the present race of mankind from maize paste. Chan-Bahlum and his father were again shown standing on either side of the tree, but this time their positions were reversed.

At right angles to the Temple of the Cross and directly opposite the Temple of the Floriated Cross was the third building in the group: the Temple of the Sun. It too had a central plaque inside an inner sanctum but this one was significantly different from the other two. While the two kings, large-sized Chan-Bahlum and small-sized Pacal, were again shown on the right and left, there was no tree between them. Instead, at the centre of the picture was a war-shield bearing the face of a god. Schele identifies this god as 'GIII', one of the Palenque trio and generally regarded as a sun god. Behind the shield, like arms displayed on the walls of some Scottish baronial castle, were two crossed spears with

Figure 26: A plaque from the Temple of the Sun (after del Castillo)

flint heads. Underneath was a relief of a throne supported on the heads of two aged gods. They were seated cross-legged, their heads bowed forwards, and one of them was clearly 'L', or Itzamna.

The full meaning implicit in the symbolism of the plaques contained in the three temples is a matter of debate and scholars are not in agreement concerning every detail. However, Schele's theory (now accepted by most Mayanologists) is that they were constructed by Chan-Bahlum as part of a propaganda campaign intended to bolster his position as ahau, or king.

His problem was one of male succession. While he was clearly the son of his father, Pacal, his paternal grandfather was not of the royal blood. Both he and Pacal drew their legitimacy through the female line: Pacal's mother, Lady Zac-Kuk. While she was the daughter of a king and, in the absence of living brothers, could claim legitimacy as a queen, the same could not be said of her son Pacal because his father was from a different, non-royal lineage. This must have put a question mark over his legitimacy as king, at least until such time as he had established himself as a capable ruler. Steps had evidently been taken to quiet the opposition and to make sure that the succession of his own son, Chan-Bahlum, went as smoothly as possible. The way this was done was ingenious to say the least. To get around the problem of matrilineal succession in a patriarchal society, Pacal and Chan linked their claims to religious mythology. According to Schele, the inscriptions on the plaques of the Temple of the Cross were intended to depict Pacal and Chan-Bahlum as acting in accordance with the will of the gods.

The inscriptions name 'Lady Beastie', the 'First Mother', as the goddess who gave rise to the Maya and say that she was born on 7 December 3121 BC. Her consort, 'First Father', was born the year before: on 16 June 3122 BC. Some eight years after his birth, on 13 August 3114 BC, the thirteenth baktun of the third age came to its end and the present fourth age began. Two years after this, First Father, whom Mayanists call god GI', ascended into the sky.

According to the inscriptions, at age 815, Lady Beastie gave birth to the three principal gods of Palenque: GI, GII and GIII. Chan-Bahlum used this story to illustrate how his own grandmother, Lady Zac-Kuk, who ruled as queen of Palenque while Pacal was a young boy, had a career that paralleled First Mother. This being so, Pacal's (and hence his own) legitimacy as king of Palenque was not to be questioned. The gap in the patrilineage was simply following a divine prototype.[12]

That Schele and the other members of the *Mesa Redonda* were able to get this far and to successfully translate the hieroglyphic inscriptions on the walls of the temples of Palenque was a remarkable achievement. However, her work did not stop there. As significant, although more controversial, were her ideas on archaeo-astronomy, which, as we will see, provide clues to a deeper understanding of not only the Palenque inscriptions but also the whole philosophy of the Maya that gave rise to their prophecies for the end of the age.

CHAPTER 7

Astronomy and the Death of Seven Macaw

One of the most extraordinary discoveries of the past few decades has been the steady realisation that much of what the Mayans and other Central Americans recorded in the form of myths relates to astronomy. Again, Linda Schele, late Professor of Art at the University of Austin, was the undoubted driving force behind this line of research. As we have seen, Schele, who died in 1997, was one of a select group of scholars who, working together in the 1970s and '80s, succeeded in translating many of the texts left behind by this lost civilisation. With some reservations, in my opinion her work on the astronomical meaning behind Maya myth and ritual was more remarkable. Not only does it seem closer to the soul of the ancient Maya but it opens up a whole new way of putting the Maya prophecies for the end of the age into context.

Schele realised that certain Maya myths are represented in pictorial form on plaster panels, stelae and ceramics. By studying the customs of the modern Maya and relating them to these pictures, she was able to show that the myths depicted often had a cosmological meaning. Not only that, she discovered that they represented stories that were recorded in word form in the *Popol Vuh*: the great Mayan epic of creation, which turns out to be a veritable goldmine of information about the cosmological beliefs of the ancient Maya.

The *Popol Vuh*, which was first brought back to Europe by Brasseur de Bourbourg in 1857, had previously been regarded as of too recent origin to be of help in understanding Mayan ideas from the Classic era. However, as more and more monumental texts have been translated, it has become clear that, far from being a modern creation, it is based on traditions going back to at least the Classic era (*c.* AD 100–900) and probably to long before that. It begins with the story of how the gods created the world and all the animals. Unfortunately, the animals they had created proved incapable even of speech. This was not satisfactory from their point of view, as they longed for a being that would speak to them and sing their praises. Accordingly, the animals were condemned to serve as food for their next creation: man. Again, their first two attempts at creating sentient humans were both failures. The first men, made out of mud, just crumbled away. The second batch, made of wood, were stiff, dumb and had no memory of their origins. It was only on the third attempt, when they made men out of maize dough, that the gods were finally successful.

The *Popol Vuh* then explains how at the end of the third age – the one prior to the emergence of the present race of mankind – a strange bird-like creature called *Vucub Caquix* (Seven Macaw) sought to take the place of the Sun. Seven Macaw was boastful concerning his majesty and usurped the role of the Sun as giver of light:

> It was cloudy and twilight then on the face of the earth. There was no sun yet. Nevertheless, there was a being called Vucub-Caquix who was very proud of himself.
>
> The sky and the earth existed, but the faces of the sun and the moon were covered.
>
> And he [Vucub-Caquix] said, 'Truly, they are clear examples of those people who were drowned, and their nature is that of supernatural beings.'
>
> 'I shall now be great above all the beings created and formed. I am the sun, the light, the moon,' he

exclaimed. 'Great is my splendour. Because of me men shall walk and conquer. For my eyes are of silver, bright, resplendent as precious stones, as emeralds; my teeth shine like perfect stones, like the face of the sky. My nose shines afar like the moon, my throne is of silver, and the face of the earth is lighted when I pass before my throne. So, then, I am the moon, for all mankind. So shall it be, because I can see very far.'

So Vucub-Caquix spoke. But he was not really the sun; he was only vainglorious of his feathers and his riches. And he could see only as far as the horizon, and he could not see all over the world.[1]

In a related story, the creation of the true race of man begins with the bringing into being of two sets of 'hero twins'. The second of these pairings, *Hunahpu* and *Xbalanque*, humiliate Seven Macaw by shooting him down from his perch on top of a fruit tree.

Vucub-Caquix had a large nantze tree and he ate the fruit of it. Each day he went to the tree and climbed to the top. Hunahpu and Xbalanque had seen that this fruit was his food. And they lay in ambush at the foot of the tree, hidden among the leaves. Vucub-Caquix came straight to his meal of nantzes.

Instantly he was injured by a discharge from Hun-Hunahpu's blowgun which struck him squarely in the jaw, and screaming, he fell straight to earth from the treetop.[2]

After he is shot from the tree, Seven Macaw suffers from toothache. To ease the pain, he begs the twins to pull his teeth, but as the teeth are drawn, so he loses his greatness. Humbled and no longer proud of his beauty, he dies. His wife, called *Chimalmat* (a word thought by Tedlock to mean 'shield stars'), also dies. They are survived by their two sons. The eldest of these is a giant called *Zipacna*, whose name is said to derive from a mythological animal with the features of a crocodile. The other, called *Cabracan*, or earthquake, is a giant who breaks mountains. Following their slaying of Seven Macaw, the boys have further adventures hunting these two giants. This brings to an end the third age and clears the way for the fourth: the Age of the Jaguar.

Having described the creation of the human race in generalised terms and the slaying of Vucub Caquix and his sons at the transition from the third to the fourth ages, the *Popol Vuh* backtracks to explain how Hunahpu and Xbalanque were themselves born. The story begins with a previous generation of twins called, confusingly, One Hunahpu and Seven Hunahpu. They are the sons of an ancient couple, *Xpiyacoc* and *Xmucane*, so old that they are more like grandparents than parents. One Hunahpu and Seven Hunahpu's favourite activity is playing a ball game. Unfortunately, the noise made by their rubber ball as it bounces on the ground so irritates the gods of Xibalba, the underworld, that they are summoned there. Here they are tricked and murdered, the head of One Hunahpu being hung from a calabash tree. This head is visited by a girl, the daughter of one of the underworld gods, who, disobeying her father's orders, goes to the tree to eat its fruit. The head talks to her and she becomes pregnant after it spits in her hand. As a result of this strange intercourse, she gives birth to the second pair of twins: the heroes who later shoot Seven Macaw. Called Hunahpu (One Lord) and Xbalanque (Small Jaguar Sun), they turn out to be much smarter than their father and uncle.

Following on from their adventures with Seven Macaw and his sons, the hero twins take up a challenge from the Xibalbans that they too should play the ball game. However, perhaps because their mother was herself a Xibalban, they are a lot wiser than their predecessors in dealing with these demon lords. By means of trickery, they pass the set of tests that their father and uncle failed, even winning the ball game. However, unknown to the Xibalbans, they have one further trick up their sleeves: they have the power to resurrect themselves after death. Knowing this, they eventually allow themselves to die by jumping into an oven. The seemingly victorious underworld gods are delighted at this. They grind up their bones and throw the resultant maize-like flour into a river. However, even this is not the end of the story as, using their magical powers, the boys resurrect themselves. They then trick the leaders of the

Xibalbans into allowing themselves to be killed with the promise that they too will be resurrected. Once the gods are dead, though, they leave them in that condition and do not restore them to life.

The overthrow of the evil Xibalbans accomplished, the boys seek their father's grave. Told that First Father lies buried in the centre of the ball-court, they use their magic powers to bring him back to life. Unfortunately, he suffers from dementia and cannot remember more than a few words. Accordingly, he has to remain in the ball-court. He is, however, transformed into the maize god, in which form he will receive praise from all future generations of people. Their work done, Hunahpu and Xbalanque ascend to heaven, where they become the lords of the Sun and Moon respectively.

This, in essence, is the central story of the creation of the present age as expounded in the *Popol Vuh*. Linda Schele had some brilliant insights concerning the cosmological meaning of these myths. These came to her after David Freidel, later to co-author her best-known book, *A Forest of Kings*, made the casual remark to her that the scorpion standing by a tree depicted on a ceramic pot represented the constellation of Scorpio (see Figure 19). He also suggested that the bird was intended to represent Seven Macaw.[3] This would imply that the story of the shooting of Seven Macaw, whose death signals the end of the third age, was to be understood as an astronomical allegory.[4]

As we have seen, inspired by Freidel's observation, Schele looked for other correspondences that might link the images depicted on this pot with astronomy. Thinking laterally, she had conjectured that if the scorpion on the pot was intended to represent the constellation of Scorpio, then maybe the tree in which the bird was roosting symbolised the Milky Way itself. This idea now led her into a deep and not unrewarding study of Milky Way astronomy, which she now believed was behind many of the Maya myths. Working with sky-maps, she soon discovered that in the course of the night the Milky Way undergoes transformations in its position relative to the horizon. Its position in the sky at any given time of day varies according to season, but in May it lies flat around the horizon at dusk. Then, as the night progresses, it gradually rises up. At midnight, it forms an arch across the sky before subsiding again to form a ring around the horizon at dawn. If the Milky Way was symbolised by a tree, Schele wondered, could the raising up of what the Maya called the wakah-chan, or 'raised up sky' tree, have been what was symbolised on the lid of Pacal's tomb? Might not the tree that looms over him as he seemingly falls backwards into the jaws of death have been intended to represent the Milky Way, or the white road along which he would travel on his way to heaven?

Linda Schele certainly believed this to be the case, even though these ideas were quite at odds with other traditions

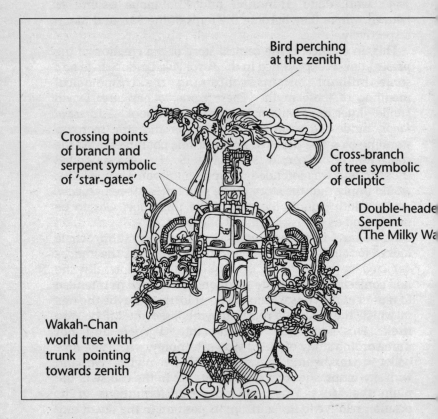

Figure 27: Lid of Palenque with 'raised up sky' tree interpreted as the Milky Way

concerning the world tree which she had earlier subscribed to. Her ideas, however, fitted quite well with what is recorded by the world's first Mayanist: Bishop Landa of Yucatán.

THE QUADRIPARTITE UNIVERSE OF THE MAYA

In his book *Relación de las Cosas de Yucatán*, c. 1570, Friar (later Bishop) Diego de Landa describes in detail the New Year festivals held in the Yucatán in his own time. These bear a startling resemblance to those still held by Maya today, proper orientation towards the four quarters still being considered of the utmost importance. Linda Schele discovered this when, in 1989, she and David Freidel attended a Mayan rain ceremony in Yaxuna in the heart of the Yucatán. Before the ceremony proper could take place, it was first considered necessary for the elders to construct an altar out of poles and vines. This makeshift structure was oriented towards the four compass directions so that a proper invocation of the rain gods could take place. Essentially, the same philosophy that underpins the building of such a Maya shrine today was, in Classic times, applied to the construction of temples. Back then, the Maya considered it of utmost importance that 'spirit houses' were oriented towards the four quarters. Indeed, Bishop Landa tells us that in Post-Classic times they oriented whole cities this way, reconsecrating them each year with special ceremonies. The four directions – north, south, east and west – were ruled over by four bacabs, or rather *bacabob*.[5] Idols symbolising them were placed by the gateways of Yucatecan cities, whose main avenues were also orientated towards the cardinal directions.

Other sources tell us that the Maya conceived the Earth as being like a two-headed turtle, swimming in the sea of space. According to these sources, the bacabob stood on the turtle's back and supported the canopy of the sky at its corners. From the centre of the turtle's carapace rose a gigantic tree, which formed a central axis. This tree was associated with the colour turquoise or blue-green and was surmounted by the *kukul* bird, known as the quetzal by the Aztecs, whose iridescent turquoise tail-feathers were much prized by Maya and Aztecs alike.

The Aztecs, as mentioned earlier, used to raise a tall pole

on their New Year feast of Xocotlvetzi. The idea of celebrating the New Year by raising a symbol of the world-tree is still alive in parts of Mexico, though these days it is symbolised by a large cross rather than a pole. Pre-Columbian tradition also lies behind the amazing dance or ritual known as 'Aztec flying', which, as described earlier, involves five men 'flying' to the ground in thirteen circles from the top of a large pole.

This Aztec/Totonac survival is clearly closely linked with the same groundwork of beliefs underpinning the Mayan idea of the wakah-chan. The flying 'bird-men' echo the Mayan teachings concerning the bacabob of the four directions, while the drummer at the top of the pole is like the kukul or quetzal who sits on top of the world tree. The 13 turns of the ropes are also of significance. The number 13 was sacred to all the peoples of Central America and generates the sacred cycle of 260 days known to the Maya as the Tzolkin. It is also the number of weeks in a quarter. As 4 x 13 = 52 weeks, the flight of the four flyers symbolises the completion of a year: the time taken for the passage of four seasons. The climb back up to the top of the pole inaugurates the beginning of a new year, which for the Aztecs and Maya took place between mid-July and early August depending on the latitude at which they lived.

Linda Schele was evidently aware of these ideas at least at the time she and David Freidel were writing *A Forest of Kings*. On page 67, they placed a diagram showing the sort of quadripartite arrangement of bacabob that Landa had described. At the centre of this diagram is a tree that is labelled 'Wakah-Chan, the World Tree'. It is depicted exactly like the one on the central panel of the Temple of the Cross at Palenque. As in the original, the tree has an elaborate bird, similar to the one on the Lid, perched on top of it. This they labelled 'The bird of the centre axis' (see Figure 28).

Looking at this diagram, there is little doubt that Schele and Freidel originally thought that the tree was to be looked upon as rising vertically from the ground – rather like the pole that was climbed by Aztec boys on the summer feast of Xocotlvetzi, which Bernadino Sahagun says was held near the beginning of August. However, by the time she came to write *Maya Cosmos*, published three years later in 1993,

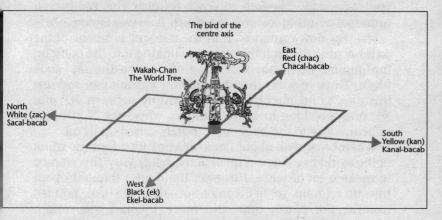

Figure 28: The world tree at the centre of the four bacabob

Schele's thinking had clearly changed. Taking note that the Maya partitioned space in eight directions – north, north-east, east, south-east, south, south-west, west and north-west, she observed that they thought of the heavens as being like a house with eight such partitions, referring to it as *yotot xaman*, or house of the north. The northerly direction is, of course, of special importance because this is the direction of the pole around which the stars appear to turn. Schele concluded from this that the Maya's axis of creation – symbolised by the wakah-chan tree – was not seen as rising

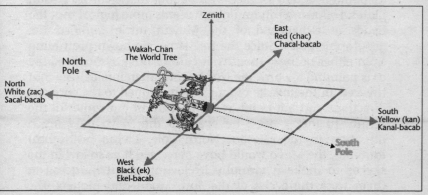

Figure 29: 'Raised up sky' tree pointing at the North Pole

vertically and therefore directed towards the zenith but rather as pointing towards the North Pole (see Figure 29).

She was now convinced that the tree was symbolic of the axis running through this pole and linking it to the point on the ground where the ceremony of 'raising the sky' took place. Thus, in ways that are difficult to visualise, she saw this axis as inclined from the vertical by between 70° and 74° and therefore, in Maya latitudes where the North Pole appears close to the horizon, as barely raised up at all.

When I first read about Linda Schele's idea that the trunk of the world tree was a symbol of the Milky Way, this seemed to make a lot of sense. However, the more I thought about this in relation to her contention that the tree pointed towards the North Star, the more I could see certain weaknesses in her argument. I could not reconcile her assertion that the bird sitting on top of the tree was meant to symbolise Seven Macaw, identified as he has been with the Big Dipper constellation. Now it is true that the Big Dipper circles the North Pole and that, as the line linking the poles is the axis around which the world turns, this could in theory be symbolised by a tree similar to the one used by the Aztec flyers. However, to judge from their art, this does not seem to be how the ancient Maya viewed the situation. As has been noted, their symbol for the Earth was a giant turtle with two heads, which is actually quite an appropriate analogy. The shell of a turtle, like that of a tortoise, is made up of flat irregular segments analogous to the Earth's crustal plates. Following this analogy, it seems more logical that the heads at either end of the Mayan turtle signified the geographic poles. Since the tree is never shown protruding from either of these mouths, it cannot have symbolised the axis pointing towards the North Star. I therefore believe that the tree, in the middle of four others aligned to the cardinal directions, did not point at the pole at all but rather at the zenith. In fact, I don't believe that the Maya paid much attention to the celestial North Pole. Living in tropical latitudes, the Maya would have perceived it as so low in the sky as to make it virtually irrelevant. What mattered to them were the cycles of the Sun, Moon and planets, the rising and setting of stars and, above all, what was culminating at the zenith.

Schele's claim that the Milky Way was the 'raised up tree of the north' also made no sense. Because of the way the plane of our galaxy is angled to the ecliptic, it does not – and never has – risen into a position that runs north–south in the sky. What in fact happens is that it twists around so that first of all one half of it rises to form an arch running more or less south-west to north-east and then, as the Earth turns, this half of it sets and the other half rises to form an arch running from south-east to north-west. Thus, to link the movement of the Milky Way with the 'world tree of the north' is clearly wrong, either in terminology or astronomy. As for the Big Dipper, it too does not and never has occupied the position of the North Pole and nor are its stars part of the Milky Way.

As I couldn't see why Schele was so convinced that Seven Macaw symbolised the Big Dipper, I decided to investigate the matter further. I discovered that Schele based this claim on work attributed to Dennis Tedlock, the translator of what is today considered the definitive English language edition of the *Popol Vuh*. The reference that Schele gave in a footnote in *Maya Cosmos* was to page 360 of Tedlock's 1985 edition of the *Popol Vuh*.

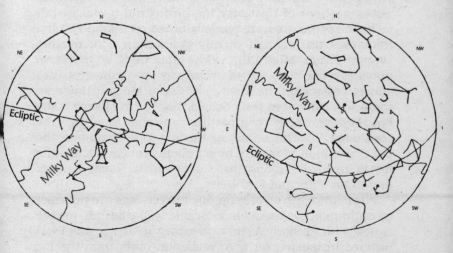

Figure 30: Two positions of the Milky Way when it arches over the sky

As this just happened to be in my own library, I decided to look up for myself what he had to say. It turned out that this was not quite the authoritative source that Schele had claimed. Tedlock admitted that the *Popol Vuh* itself did not specify which constellation (if any) represented Seven Macaw. However, he said that one of his sources, given the initial 'A', indicated that it was Ursa Major because its brightest stars (those which make up the Big Dipper or Plough constellation) were seven in number.

After reading this, I decided to look up the identity of the mysterious 'A' who was Tedlock's source for identifying Seven Macaw with the stars of Ursa Major. He turned out to be Miguel Alvarado López, author of a Quiché-Spanish medical dictionary. As this seemed to me to be rather a weak authority on which to base a whole archaeo-astronomical theory of correspondences, I decided to see what others had to say on the subject. Rather surprisingly, there was nothing about Seven Macaw being the Big Dipper in Anthony Aveni's now classic text on Central American astronomy, *Skywatchers of Ancient Mexico*. Karle Taube, in his book *Aztec and Maya Myths*, narrates the story of Vucub Caquix but points out that in ceramic representations of this story the bird is not a macaw but rather a mythic creature possibly based on a king vulture. He makes no mention of any connection with the Big Dipper. Susan Milbraith, whose *Star Gods of the Maya* provides a useful compendium of competing theories, also quotes Tedlock on the identity of Seven Macaw. However, she is less convinced than Schele that Seven Macaw is the Big Dipper. After contrasting various theories, she concludes: 'Identifications of the Big Dipper among the pre-Columbian Maya are by no means conclusive, despite recent research linking this star-group to Vucub Caquix and to the Principal Bird Deity.[6]

Susan Milbraith's words on this subject – she is a professor of anthropology, knew Linda Schele well and has worked closely with Anthony Aveni – gave me confidence that I was justified in looking for a constellation other than the Big Dipper as representative of Vucub Caquix. After considering all of the evidence, I came to the conclusion that the world-tree was vertical and pointed towards the zenith. This meant that Vucub Caquix, Seven Macaw, must have 'sat' at the zenith. This made sense as the Maya, like the Aztecs, were much more interested in the zenith than the pole and it is this point in the sky, which stands directly overhead, that they still call *y hol gloria*, or the glory hole.

Further research revealed that the zenith point was looked upon as being some sort of portal to a nether-world. It was both the 'raised up heart of the sky' (*Wak-chan-ki*) and a sort of birth canal from which emerged umbilical cords. Although, because the sky rotates, certain points on the Milky Way will cross this 'hole' at the zenith, in Mesoamerica the Big Dipper (known in the United Kingdom as the Plough), which is not even on the Milky Way, never does so. Not only that, but this constellation is quite far removed from the Milky Way. Thus, in my opinion, it is difficult to see how the Big Dipper could ever have been visualised as a bird sitting on the wacah-chan if this be equated with the Milky Way.

After carefully considering this problem, I now believe it to be nothing more than a misfit of pieces of the Mayan jigsaw. Once it is accepted that Seven Macaw was not originally thought of as symbolising the Big Dipper constellation, then the pieces fit together properly and the problems melt away. I shall return to this shortly.

The reason the Maya were interested in the zenith was twofold. First of all, for anyone living in the tropics it is noticeable that there are two days in the year when the Sun will cross this point at midday and a vertical pillar will cast no shadow. Which days these are depends, of course, on the latitude of the location: the further north (and therefore nearer to the Tropic of Cancer), the nearer these days will be to the summer solstice. On the Tropic of Cancer itself, the Sun will perform this magic on only one day: exactly on the summer solstice. Conversely, at places on the equator, the Sun will transit the zenith exactly on the equinoxes. The Maya lands extend from about 14° 30' N in South Guatemala to 21° N in the northern tip of Yucatán. Different Mayan cities would therefore have witnessed the second solar transit on different days depending on latitude, the variation being between 16 July and 13 August.

ASTRONOMY AND THE FALL OF SEVEN MACAW

The interest in zenith transits of the Maya (and other tropical peoples such as the Aztecs, Zapotecs and Olmecs) sits uncomfortably with the concept that Seven Macaw symbolised the Big Dipper. As stated above, I now believe that this identification is wrong. As we have seen, the foundation myth of the Aztecs concerned a vision of an eagle grappling with a serpent. This story, which is in the nature of a fable, is illustrated in the *Codex Mendoza*, an early post-colonial book. Here there is a picture of the eagle sitting on top of a Saguaro cactus, with a serpent climbing the rock on which it is growing. In other books, the serpent is shown hanging from the eagle's beak. The eagle grappling with a serpent is still the prime symbol of the Mexican nation and is today to be found at the centre of the flag of Mexico. This is for a good reason. Close inspection of the image in the *Codex Mendoza* reveals that the cactus is the Aztec equivalent of the Mayan tree of life. Around it is a square arrangement divided into four segments. This would suggest that the eagle is an analogue of the Mayan Seven Macaw. That the story of the shooting of Seven Macaw was not a Mayan invention is indicated by his presence on monuments at Izapa, a pre-Mayan city near to where the earliest long-count dates have been found. The question now is: if this bird did not symbolise the Big Dipper, what did it represent?

One of the charges laid against Seven Macaw is that he sought to take the place of the Sun. He boasts, 'I am great. My place is now higher than that of the human work, the human design. I am their sun and I am their light, and I am also their months . . .'. These words annoyed the hero twins, who '. . . saw evil in his attempt at self-magnification before the Heart of the Sky'. Now, as we have seen above, the 'Heart of the Sky' is the zenith point and not the North Pole around which the Big Dipper rotates. What, then, could be the meaning of these words?

I found an answer to this question by consulting my computer – more specifically, a program called *Starry Night*. This program is capable of recreating how the sky looked throughout the era of the Maya's fourth age, i.e. from 3114 BC to AD 2012. Taking into account the movement of the

Figure 31: Cover of the Codex Mendoza

stars caused by the 'precession of the equinoxes', it is able to show which stars were rising and which culminating at any place on Earth. Remembering the episode of Hunahpu hunting Seven Macaw with his blow-gun and how Linda Schele had identified the scorpion often depicted with this incident as being the constellation Scorpio, I looked first at this part of the sky. Immediately, it was clear what this episode was all about. Next to Scorpio is the constellation of Sagittarius, the archer. Perhaps this represented Hunahpu. Higher on the 'tree' of the Milky Way was the constellation of Aquila: Latin for eagle. Like the Aztec eagle, Aquila is shown reaching out to a snake: the long, slithery constellation of *Serpens*. What was even more extraordinary was that in European mythology Aquila too is in his death-throes. He has in fact been shot with an arrow and in many medieval maps of the heavens the constellation of *Sagitta* (the 'arrow') is shown piercing his wing. In European mythology, this arrow is dispatched by the hero Hercules, whose constellation lies nearby and who, like Hunahpu, is one of a pair of hero twins (I will return to this later). The clincher as far as I was concerned was that Seven Macaw's wife was called *Chimalmat*, a name which Tedlock says meant 'shield'. His further identification of her as the constellation of the Little Bear seemed contrived to fit with the, in my opinion, erroneous idea that Seven Macaw was the Big Dipper. However, it seemed more than coincidental, to me anyway, that on European star-maps is shown, just below Aquila, the constellation of *Scutum*: Latin for shield.

Next to Aquila and close by his feet is the tail of the constellation of Serpens. Below this, a little to the north of Scorpio's sting, is another extremely significant point in the sky. This is where the ecliptic, the pathway followed by the Sun, crosses over the median plane of the Milky Way. This celestial coordinate, which I referred to in *The Mayan Prophecies* as the southern, or Scorpio, 'star-gate', is the counterpart of another star-gate which, because it is very important from the point of view of biblical prophecy, I wrote about extensively in my book *Signs in the Sky*. However, the Scorpio star-gate is also highly important, as it coincides with the very centre of our galaxy.

Taking all these matters into consideration, it seemed likely to me that the Lid of Palenque was meant to show King Pacal falling backwards and about to enter the afterlife through this gateway. The tree which rises above him represents the great ceiba – world tree – that symbolically rises to the zenith point. However, I do not believe that wakah-chan means 'raised up sky' but rather 'raised up serpent': the word chan (or *kan/can*) being homophonous and having a dual meaning in the Mayan languages as both 'sky' and 'serpent'. This 'raised up serpent' is not hard to find. It is the serpent that is draped over the topmost part

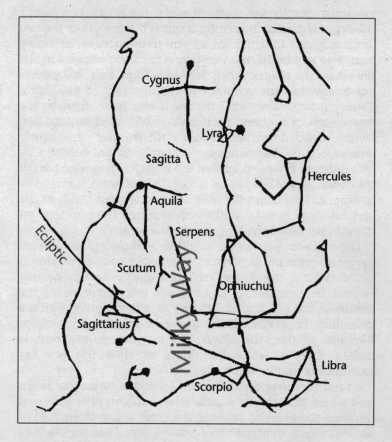

Figure 32: *Astronomy print of part of the sky with Aquila, Serpens, Sagitta, etc.*

of the tree on the lid of Pacal's sarcophagus and which Linda Schele believed represented the ecliptic (see Figure 33).

Again, I found I questioned her reasoning in associating this serpent with the ecliptic. It seemed much more likely to me that it symbolised the Milky Way itself. This would explain the curious knobbles along the body of the serpent. Elsewhere in Mayan and Aztec art, such knobbles symbolised stars. Of course, the ecliptic also carries stars, but nowhere near as many as the Milky Way. It seemed to me that having the serpent held aloft by the tree was intended to mirror exactly the moment when the Milky Way spans the sky and crosses the zenith. If this is the case, then the two 'dragon heads' in which the serpent terminates are probably meant to symbolise the star-gates: i.e. the positions in the sky where the ecliptic meets the Milky Way. This, however, is not to say that the ecliptic was unimportant to the Maya. Clearly it was. However, I believe it was symbolised by the cross-beam of the tree: itself a double-headed serpent bar. Double-headed serpent bars, which are frequently represented on bas-reliefs and sculptures, are believed by Mayanists to have symbolised the ecliptic. They were part of the royal regalia, rather as a sceptre is a symbol of royalty in Europe. This being the case, the two serpent heads at the ends of these bars very likely represented the positions on the ecliptic where the Sun transits the zenith.

These ideas led me to a subtle redrawing of Schele's diagram, restoring the tree to an upright position so that it could support the Milky Way as Aquila (Seven Macaw) crossed the zenith. As regards my interpretation of the design on the pot, the bird clearly signifies Aquila, with the spreading branches of the tree (not the trunk) perhaps symbolic of the Milky Way. The scorpion at the base is indeed Scorpio, while the figure of Hunahpu possibly signifies the Sun in Sagittarius.

To test this hypothesis further, I decided to look at when and where the stars of Aquila were transiting the zenith at the time the Mayans believed the present age started: 3114 BC. This was not difficult to do as the *Popol Vuh* says that at the start of the present age people gathered together in darkness at a place called *Tulun Zuyua*, or Seven Caves, Seven Canyons. The location of this 'Tulun' has in the past been a matter of debate but there seems little doubt that, in Classic times at least, there was a general consensus throughout Mesoamerica that it was Teotihuacan. This was the largest city in America, had a network of seven caves concealed beneath its largest pyramid and, like some Mesoamerican Jerusalem, drew pilgrims from far and wide.

Accepting for now the premise that Teotihuacan was the original Tulun Zuyua, I set up my computer program to show how the stars would have appeared there in 3114 BC. I then used the program to 'move' the sky hour by hour and

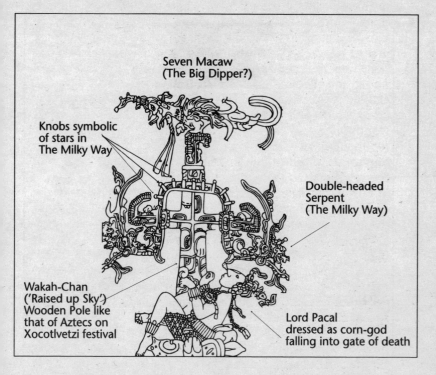

Seven Macaw
(The Big Dipper?)

Knobs symbolic
of stars in
The Milky Way

Double-headed
Serpent
(The Milky Way)

Wakah-Chan
('Raised up Sky')
Wooden Pole like
that of Aztecs on
Xocotlvetzi festival

Lord Pacal
dressed as corn-god
falling into gate of death

Figure 33: Tree and serpent as the Milky Way held up at the zenith

minute by minute. To my surprise, I found that on 13 August, the start-day of the present cycle, the topmost 'wounded wing' of the constellation of Aquila would have been seen to transit the zenith just after sunset. It would be followed almost immediately by the little constellation of Sagitta – the arrow – which in Western mythology is fired at Aquila by Hercules.

This could be equated with the blow-dart shot at Vucub Caquix by Hunahpu. Thus, in a very graphic way, the stars over Teotihuacan told the same myth as the *Popol Vuh*: that our present age began with the displacement of Aquila from its 'usurped' position of transit across the zenith.

Intrigued by this discovery, I now used the same program to see when the constellation of Aquila began transiting the

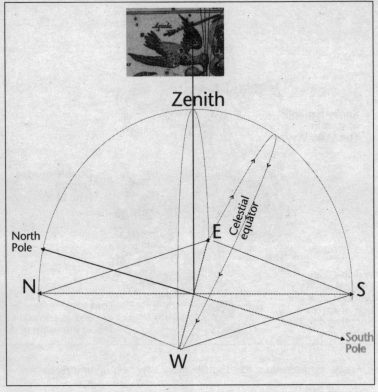

Figure 34: Wounded wing of Aquila transiting the zenith in 3114 BC

zenith. I discovered that its most southerly star, *Theta Aquilae*, which corresponds to the head of the bird, was transiting the zenith at around 7395 BC, i.e. 4,281 years before 3114 BC. This was in fairly good agreement with the Aztec account contained in the *Vatico Latin Codex*, which says the previous age lasted for 4,081 years.

What all this seemed to be saying was that the third age – the one prior to our own – corresponded to the time when Aquila was transiting the zenith over Teotihuacan. Uncannily, not only did these much later people of the fourth age seem to have remembered that Teotihuacan was a place of extreme importance, but both Maya and Aztec myths linked its foundations with a bird deity we would

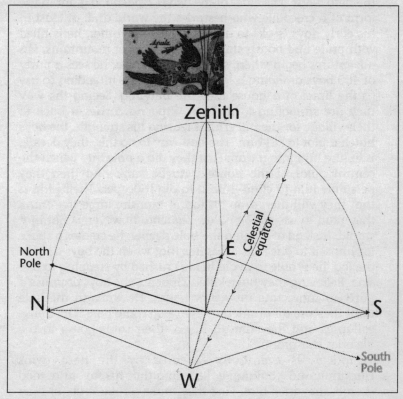

Zenith

North
Pole

E Celestial equator

N

S

W

South
Pole

Figure 35: Head of Aquila transiting the zenith in c. 7395 BC

recognise as personified by the constellation of Aquila. This, however, was not the end of the story concerning zenith calendrics and Teotihuacan. I was now to find there was yet more astronomy connecting the myth concerning Seven Macaw's sons and the destruction of Teotihuacan in the eighth century AD.

THE SONS OF SEVEN MACAW AND THE TEOTIHUACAN ZENITH

The story of Seven Macaw does not finish with his death. He leaves behind him two sons, Zipacna and Cabracan, who have their own adventures before they are finally slain by the hero twins. The elder of the two is Zipacna, a giant in the form of a crocodile who bestrides the world and, at least in his early days, seeks to do good. Like his father, he is filled with pride and boasts that he is a maker of mountains. His adventures begin when, sitting by the coast, he sees a party of 400 boys dragging a log which they are intending to use as the lintel of a house they are building. Seeing the way they are struggling to move it, Zipacna carries it back to their village for them. Perhaps fearing his strength, the boys hatch a plot to kill him. The best way to do this, they decide, is to lure him into a trap. First they dig a hole into which the central pillar of the house is to be sunk and then they persuade him to climb into it to dig it deeper. Their plan is that they will then drop on top of him the large tree-trunk they plan to use as the pillar. Zipacna, however, is not that stupid. Instead of digging the hole deeper, he creates a space for himself to one side. Thus it is that when the boys drop in the log, he is able to avoid being crushed by stepping to one side. Believing they have killed Zipacna, the boys organise a party to consecrate their new house. He waits in the hole until they are all drunk and then hoists the pole. The house collapses and they are all killed, their souls going to the Pleiades star-group.

Zipacna is eventually defeated by the hero twins Hunahpu and Xbalanque. Learning that his favourite food is crab, they trick him. First they make a fake 'crab' of their own out of some bromelia flowers and a boulder. They hide this in a cave at the base of a ravine and then persuade Zipacna to go in after it. He does so and becomes trapped in the cave, turning into solid rock. Leaving him there, they turn their attention to his younger brother, who, like the rest of his family, has an inflated view of his own importance. Whereas Zipacna boasted that he could create mountains, Cabracan's claim to fame is that he flattens them. The twins shoot birds with their blow-guns and make fire using a drill. They roast one for Cabracan, carefully coating it with earth. He eats it but the earthy coating takes away his strength. Taking advantage of his weakened state, the twins bind him hand and foot and bury him in the ground. This brings to an end the dynasty of Seven Macaw and opens the way for the birth of the new age.

Reading this, I felt certain that on one level it is a 'Just So' story: one intended to explain some otherwise extremely strange discovery. The story of how Zipacna, a giant crocodile-like being, was turned to stone in a cave could have a very simple explanation. Many fossils from the Cretaceous period have been found in Mexico, including those of Albertosaurus, a close relative of the more famous T. Rex. We know that at least some of the early Maya lived in caves. For instance, the limestone caves of Loltun (near Labna in the Yucatán) are extensive and are known to have been occupied by humans since at least 2000 BC and quite possibly much earlier.[7] These caves, and many others in the Yucatán, are the result of erosion caused by underground rivers. As the bones of a woolly mammoth have been found at Loltun, it is not impossible that the ancient Maya came across other fossils. Some of these could have gone back to the Cretaceous period, when the limestone rocks that make up the caves were being formed. Depending on the state of preservation, if they found the bones of an Albertosaurus, the Maya might have rightly concluded that, giant as it was, it belonged to the crocodile family. Such a discovery could have given rise to the story of the hero twins tricking Zipacna into a cave and his subsequently being turned into stone.

Curiously, although the story of Hunahpu and Xbalanque trapping saurian monsters is a myth, the role of Mexico in the mass extinction of dinosaurs is not. It is now believed that this was caused by an asteroidal impact that occurred on the coast of the Yucatán Peninsula about 65 million years ago. This really was the ending of one age – that of the dinosaurs – and the beginning of another, the age of mammals. It seems remarkable, but while the Classic Maya are unlikely to have known that the extinction of dinosaurs took place as long ago as it did, they were spot on in attributing the demise of crocodile-like creatures to the end of an entire age. Furthermore, the family link between Zipacna and a bird-like creature with teeth (Seven Macaw) is also clear from the dinosaur record. No birds today have teeth but bones of pterodactyls, the flying dinosaurs of the Cretaceous period, have been found in Mexico. Might not an intuitive memory of these giants of the air have given rise to the story of Seven Macaw? Alternatively, might there once have been a Mexican 'Lost World' where early humans came up against pterodactyls and shot them from trees? Such a possibility seems unlikely, especially since the impact of the asteroid thought to have caused the extinction of the dinosaurs happened close by in the Yucatán. However, while it seems to me more likely that the Maya formed their mythology to explain the bones of extinct dinosaurs rather than the living, we cannot dismiss the possibility of contact completely. Though we may never know the answer for sure, humility alone demands that we keep an open mind concerning the possibility of contact. Perhaps dinosaurs weren't quite extinct in 3114 BC, when the Maya claim our present age began. That might at least explain the fascination by cultures east and west with dragons. They, after all, look a lot like dinosaurs and some of them fly, too.

These deliberations aside, I was conscious that there was probably more to the story of Zipacna than a race-memory of human conflict with the last of the dinosaurs. Since the story of Seven Macaw clearly had meaning in terms of astronomy, I was fairly sure the same would be true for Zipacna. The story is very interesting from the point of view of zenith astronomy. According to Schele and other Mayanists, in addition to being described in terms of the 'raised up sky' tree, or wacah-chan, the Milky Way was also visualised as a giant upside-down crocodile. This crocodilian version of the Milky Way is depicted on the last page of the most famous Mayan codex: the *Dresden*.

Here the Milky Way is shown, with planetary symbols on its body, pouring forth the flood that destroyed the last age of man. The open 'jaws' of this crocodile were equated by the Maya with the dark rift at the centre of our galaxy where the light of stars is obscured by clouds of dust. This region extends upwards from the vicinity of Sagittarius, past Serpens and Aquila to end by Deneb, the northernmost star of Cygnus. The upside-down crocodile therefore refers to the same region of space as is indicated in the blow-pipe incident.

A further clue as to what is going on is that the Aztecs identified the start of the present age with a sacrifice made

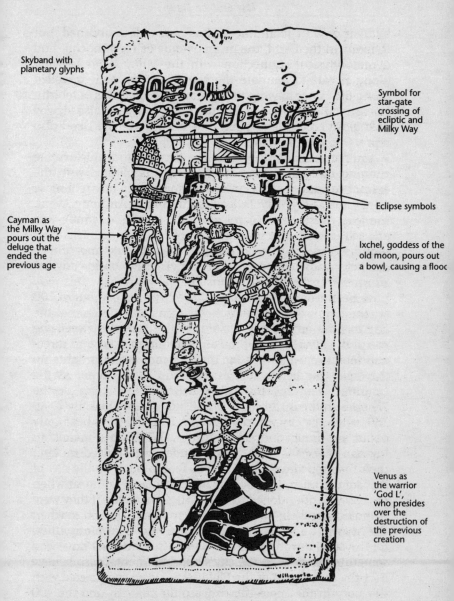

Skyband with planetary glyphs

Symbol for star-gate crossing of ecliptic and Milky Way

Eclipse symbols

Cayman as the Milky Way pours out the deluge that ended the previous age

Ixchel, goddess of the old moon, pours out a bowl, causing a flood

Venus as the warrior 'God L', who presides over the destruction of the previous creation

Figure 36: Deluge on the last page of the Dresden Codex

by their gods at Teotihuacan. It is now well established that the Way of the Dead, the main avenue of Teotihuacan, had a metaphysical connection with the Milky Way. To walk along it was to emulate the soul's journey as it travelled along that star-strewn path on its way to the gates of rebirth. However, the knowledge that the Milky Way was also viewed as a giant crocodile implied that the Way of the Dead also represented its prostrate body.

Against this, we must take into account that this avenue running through the heart of the city is aligned with the extinct volcano, Cerro Gordo, and does not in any real sense mirror the Milky Way. Indeed, according to Anthony Aveni, the east–west axis (perpendicular to the avenue) was aligned towards the setting point of the Pleiades. This star-group, he says, rose heliacally (i.e. made its first appearance just before dawn after a period of invisibility) on the day the Sun made its first zenith transit of the year.[8] Now this may have been true at some stage of the city's history, but at 100 BC, the date assigned by archaeologists to its founding, the gap in time on the day of solar zenith transit between the rise of the Pleiades and the Sun is about an hour and three-quarters. Even allowing for the fact that the Sun lights up the dawn sky before it rises, this gap seems large. As the centuries went by, the delay between the rising of the Pleiades and sunrise itself would have shortened. By *c.* AD 750, when the city was abandoned, it would have been only about an hour and ten minutes. What this means in layman's terms is that if the Pleiades were indeed making their first appearance at dawn at Teotihuacan on the day of first solar transit, they would not have been doing so when the city was abandoned in *c.* AD 750. Conversely, if they were rising heliacally in AD 750, then they weren't in 100 BC when the city was built. Either way, this must have been unsettling for the city's astronomer-priests and I suspect may have had something to do with the city's abandonment. I also believe that the movement of the Pleiades is closely connected with the story in the *Popol Vuh* concerning Zipacna and the 400 boys.

As we have seen, Friar Bernadino Sahagun, the early Spanish chronicler of Mexican traditions, recorded in his books that the Aztecs watched the movements of the

Pleiades at the end of a gavilla, or 52-year period of their calendar. Gathering on the hilltops on a day in early November with all their fires extinguished, they would watch the zenith point at midnight to see if the Pleiades stopped moving. They never did, of course, as that would mean the Earth had stopped rotating. Having confirmed this, they would then make sacrifices, light fires and generally celebrate their good fortune at being granted more time by the gods.

Aware that the derelict city of Teotihuacan was important to the Mayans as well as the Aztecs as a place intimately connected with the birth of the present age, I decided to see if there could be any connection between it and the story of the fall of Seven Macaw and his sons. Since the Pleiades is important in the later part of the myth as the place where the souls of the 400 boys are taken, I decided to look at them first. Using a computer program to backtrack the sky to how it looked in earlier epochs, I was surprised to discover that there was a further close correlation between when archaeologists say the city was abandoned and the movement of the Pleiades.

Now the Pleiades, as we see it with the naked eye, is a rather small cluster of seven stars. However, with the aid of a small telescope or even a pair of binoculars it can be seen that in actuality there are far more stars than this. There are at least 100, perhaps as many as 400: one for each of the boys. Even so, it is quite a compact group that extends to only a couple of degrees of space in any direction. Checking with the computer, I was able to ascertain that during the period of approximately AD 725 to 850 the Pleiades would have been seen to cross the zenith point – the first date relating to transits of Maia, the most northerly of the bright stars, and the later date to transits of Merope, the most southerly. A median date of AD 800 would therefore correspond to the transiting of the centre of the group. The full transit of the star-group would take only about four minutes to complete.

Now the period from AD 725 to 800 is very interesting, as archaeologists claim that Teotihuacan was abandoned around then. Because we don't have written records to tell us the reason why, it is assumed this was done for pragmatic reasons, such as warfare, environmental collapse, revolution or maybe all of the above. Whatever is the case, it seems that around AD 750, when the Pleiades were starting to transit the zenith over Teotihuacan, the city, already in decline, was ritually burnt. Archaeologists are confused as to why this was done, but it seems to me that we have only to look at the Aztec and Mayan records to understand why. We know that they counted time in calendar cycles of 52 years' duration. At the end of one of these periods, they would burn their furniture, destroy their crockery and renew their pyramids. Could it be that the Teotihuacanos did the same?

Now, while this is only a conjecture, it seems to me that the above is a distinct possibility. We know from colonial records that the last Aztec fire festival took place in November 1507. Counting back fourteen cycles brings us to a date of AD 727. If at midnight on 20 November that year the Teotihuacanos looked upwards, they would have seen Taygeta, one of the less bright stars of the Pleiades, transiting the zenith. At this moment, the Teotihuacanos would have seen the Milky Way arching across the sky, from south-east to north-west. On the western horizon would have been the top end of the black space marking the mouth of the crocodile. We therefore see depicted in the sky the scenario of Zipacna, his mouth agape, disappearing into the Earth as the world tree is held aloft on a pole, at the culmination of which are the 400 boys of the Pleiades. This visible astronomy therefore makes perfect sense of the story in the *Popol Vuh*.

Having the Pleiades transit exactly through the zenith was something that had never happened before at Teotihuacan. Yet we know the Pleiades was already an important star-group as far as the Teotihuacanos were concerned because when it was built, the Pyramid of the Sun, the largest structure in the city, was oriented in such a way that it faced the group's setting point. Seeing it transit over the zenith instead of rising heliacally with the Sun on the day of first solar zenith transit would have been an important omen for them and perhaps required that they destroy the city by fire. They then buried what remained of their city and moved elsewhere. This scenario, which I now

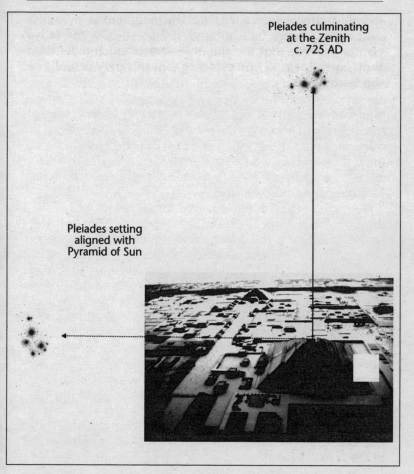

Pleiades culminating
at the Zenith
c. 725 AD

Pleiades setting
aligned with
Pyramid of Sun

Figure 37: Zenith transit of the Pleiades over Teotihuacan

believe to be the truth, does not tell us why the city rose to such importance to begin with.

Just why Teotihuacan was so important is one of the great puzzles of history. However, by putting Mayan and Aztec records together, I believe we can find an answer to this conundrum and begin to understand its significance. Teotihuacan reached its peak population of about 200,000 at around AD 600–700 and was of considerable importance to the Maya as well as the later Aztecs. However, although today's archaeologists tell us that the main plazas, pyramids and avenues were built around 100 BC–AD 300, the Maya clearly believed that the site at least was much older than that. Later, I was to find evidence that this may actually be the truth.

CHAPTER 8

Journeys from the East

Departing from the Valley of Mexico, I found myself once more in an aeroplane floating above the clouds. An hour later, we were landing in Villahermosa, one of the fastest growing cities in Mexico and the capital of its booming oil industry. This, however, was not my reason for paying a visit to what is a charmless caricature of modern urban America. I had an altogether different reason for being there than the majority of other passengers, most of them businessmen: I was going to a park.

Accompanied by other members of the tour group, I left our hotel and went by foot to what at first sight appears to be nothing more than a rather old-fashioned zoo. I had been here before so I knew what to expect, but for others in the party it came as a shock. Beyond the turnstiles were birdcages, some of which were so large that you could go inside and walk around. These cages contained tropical birds, including a pair of exotic quetzals: the sacred bird of the Maya.

The natural habitat of these birds, which are about the size of pigeons, is high-altitude cloud forest. Unfortunately, because most of their habitat has been cleared in recent years, quetzals are now much diminished in numbers and outside of zoos like this are rarely seen today in Mexico.

Fortunately, they do survive in some numbers in the relatively intact jungles of El Salvador. Even so, they can be difficult to spot unless you know what to look for. Although their plumage is iridescent, they are perfectly adapted to their lives as forest-dwellers and tend to blend in with their surroundings. As is common with birds, the male is the more colourful. With his red chest and turquoise-green back, he is quite stunning to look at. His most distinctive feature, however, is a pair of 2-ft long tail-feathers that wave in the air when he is flying. Like many birds, quetzals make their nests in holes in trees and the male and female take it in turns to incubate their eggs. However, so long are the male's tail-feathers that he cannot bring them inside the nest. When he enters the nest, he leaves them hanging outside of the hole in the tree, looking for all the world like the fronds of tree-growing ferns. To the detriment of the quetzal, not everyone was fooled, least of all the Maya, who prized these tail-feathers. They used them in their ceremonial costumes and later traded them with the Aztecs. An Aztec crown incorporating the feathers of dozens of quetzals, almost certainly given by Montezuma to Cortés, can be seen today in the British Museum.

Unfortunately, the quetzals must have been asleep on the day of our visit so we had to make do with cockatoos. Leaving them to their noisy chatter, I walked across to a pond containing terrapins and another with small alligators floating beneath its murky surface. The smell coming from them was none too pleasant and would presumably be even worse in midsummer. So too was that emanating from another cage nearby. Here, looking very much the worse for wear, was a ragged old jaguar. Clearly driven mad by captivity in a cage that could not have measured more than 20 ft x 20 ft, he stood like a shell-shocked veteran of the Battle of Verdun, moving his head from side to side. I felt sorry for him, not least because his plight was, in a curious way, emblematic of the ancient Mexico I had come to see. For if any creature exemplifies the ideals of the ancient Mayan civilisation, then it is the jaguar. A symbol of royalty, their priests often wore jaguar skins while their kings sat on jaguar thrones. The jaguar god himself was linked with both the Sun and the Moon. He was regarded as a patron of their royal dynasties. Living jaguars, like lions in Africa, were treated with respect as creatures imbued with divine power. How far away all that seemed from the situation of this old cat, locked in a tiny cage to be gawped at by tourists.

Leaving the jaguar and feeling ashamed of how humanity still treats wildlife, we left the zoo and sought inspiration in the part of the park after which it is named. Here, placed at strategic intervals along a jungle trail, is a unique collection of sculptures. They come from La Venta, one of the three major cities of the Olmec world, which is thought to date back to around 1000 BC. Unfortunately, the Olmecs built their city on top of what has since turned out to be a highly productive oilfield. The discovery of this oil in the 1970s placed the remains of La Venta under threat. Fortunately, not everyone was blind to the cultural significance of what was in danger of being lost forever. Thus it was that some 31 sculptures were taken from La Venta and moved to their present place of sanctuary in the *Parque Museo de La Venta* in Villahermosa.

It was with mixed emotions that I viewed these sculptures: sad that I was not seeing them in their correct context but nevertheless pleased that they had been saved from destruction and had been put in a park rather than a museum. At least here they could be admired in something approaching their original setting, in an environment that simulated jungle. Passing through the gate, I began by walking up the little path that led to a collection of basaltic tiles laid out in the form of a jaguar's head. At the time of the Olmecs, this probably served as a sacred space for initiation into the jaguar cult. Unfortunately, here in the park, in the presence of hundreds of tourists, all sense of mystery was stripped away. Though interesting from an archaeological point of view, it was difficult in the presence of so many flashing cameras to imagine just what rituals might have taken place upon these tiles over 2,000 years ago.

Near to the jaguar tiles was what at first appearance seemed to be a little log cabin. It was only on closer approach that it became clear that the 'logs' from which it was constructed were not made of wood at all but were actually hexagonal cylinders of solidified basalt. As each of these weighed several tons, it was clear that this building, which seems to have functioned as a tomb, represented a major achievement. The use of such sturdy building material meant that not only was it virtually indestructible, but once it was built it would have been almost impregnable to tomb-robbers, animal and human alike.

Further down the trail was the first of several gigantic sculpted heads. At least seventeen of these heads have been found, including two which are now kept in the Museum of Anthropology in Mexico City. They are free-standing monuments and their purpose, unfortunately, is unknown. Looking at the head, I was once again struck by its enormous size – over 6 ft in height. Yet the plain of Tabasco, where the Olmecs built their cities, is a swampy marsh and devoid of native stone. The basalt from which they carved these heads and other statues, not to mention the cylinders used in the impregnable tomb, must have come from elsewhere. Archaeologists believe this to have been the Tuxtla Mountains, some 150 miles to the west of La Venta, though somewhat closer to other Olmec centres such as Tres Zapotes. How, without the aid of cranes and modern transport systems, the Olmecs moved blocks of basalt weighing up to 20 tons such long distances is one of history's great enigmas. One theory is that the stones were brought down from the mountains by river and then transported by sea and up other rivers to their final resting place. If this is so, then it speaks highly of the Olmecs' navigation skills, not to mention their brute strength.

I nicknamed the head 'Baby-face' on account of its cherubic features, and this estimation of the subject's age didn't seem too wide of the mark. I looked at the way the artist had given the face a very particular characteristic. It was not a stock piece churned out according to a set formula but was clearly intended to be a portrait of a particular individual. This individual was a young man, perhaps in his late teens or early twenties. His face was round, his lips full and through his slightly opened mouth baby-like small teeth could be seen. Baby-face was not alone, as there were other, similar heads nearby. However, it was noticeable that none of them looked anything like him. Each head was evidently a portrait of someone different. The one before me might have been baby-faced but he was clearly a powerful man and not at all like the average Mayan of today. If he was a typical Olmec, then it was just about possible to imagine teams of such robust giants manhandling basalt blocks into place. However, there was something else that was strange about him, and several of the people with me commented on this. With his round head and flattened nose, he didn't look like a native American at all but rather negroid. Once more that raised the intriguing question: might the Olmecs, like the ancestors of today's black Americans, have come from Africa?

Though this is taboo among today's departments of archaeology, it is hardly new to suggest that this might be the case. It was certainly the opinion of many early explorers of Palenque that the ruins there indicated an influence from Africa. Indeed, the first person to write about Palenque, Father Ordoñez, claimed to have found a book that said Palenque was founded by people who came from across the Atlantic. In his own book, *A History of the Creation of Heaven and Earth*, he says their leader was called Votan and he originally came from a city called *Valum Chivim*. Though his report is generally dismissed today as a fictional invention, it would at least explain some of the curious similarities between the Mayan civilisation and the culture of the ancient world. If he was right and people had come to America from Africa, then it would be surprising if none of these immigrants were black-skinned. That would explain why some of the Olmec heads, including the one I was standing in front of, have what are generally regarded as typically negroid features.

More recently, there have been a number of research papers published on the Internet presenting evidence that the Olmecs were Mandeans, descended from immigrants who came to America from West Africa. It is claimed that Olmec inscriptions, which are mostly found on small objects such as celts (axe-like implements), are written in a dialect of Mandean. The evidence for such a migration seems strong to me but there is still more work to be done before this theory will receive general recognition.

The people who sculpted the heads did not call themselves 'Olmec' but rather *Xi*. In fact, the name Olmec means 'rubber people' and was given to the inhabitants of the Tabasco region by the Aztecs. They called them Olmecs because Tabasco was where the trees grew from which rubber-latex was obtained. Since rubber balls as well as the remains of primitive ball-courts have been found at several Olmec sites, it is clear that latex technology is very ancient. It is assumed that it was the Olmecs who invented the ubiquitous ball-game that played a big part in the life and ceremonies of most later Mesoamerican cultures. As the giant heads wear what appear to be sports helmets, it has also been suggested that they represent famous players of the game. Whatever is the truth of the matter, there can be no doubting the skill of the Olmecs as sculptors. Basalt is an exceedingly difficult material to work with and making these heads must have been very hard work.

In 1999, I had the opportunity of discussing these matters with Professor Glynn Williams, head of the sculpture department of the Royal College of Art in London. With works on display throughout the world, he is not only one of the leading sculptors of our times but an expert on the history of the art. In 1998, he had been invited by the BBC to take part in a programme concerning the sculpture of the Olmecs. The premise of the programme was simple: a team of volunteers would tow a large block of basalt for 100 yards and then, using Stone Age tools, Professor Williams would sculpt it into something similar to an Olmec head. However, while this seemed simple enough, doing this turned out to be much more difficult than anticipated. Although two weeks of chipping away with Stone Age tools left Professor Williams exhausted, it made little impression on the block of basalt.

It was some weeks after his return to London that he and I met in the congenial surroundings of the Royal College's canteen. Over lunch, he told me about his experiences with the BBC programme and how he had been privileged to see two new sculptures that had only just been discovered. As he had taken photographs of them, he was able to show me what they looked like prior to the broadcasting of the programme. Again, they were made of basalt but were of kneeling Olmec priests rather than the more familiar heads. The statues were in such good condition that they looked as though they had been carved yesterday and not over two and a half thousand years ago. Shown wearing jaguar headdresses, they were beautifully proportioned and evidently smooth to the touch. However, what interested Professor Williams most was not just the statues themselves but what they represented in terms of art. He explained this to me in layman's terms.

'Two weeks of hard work using what we take to have been the sort of Stone Age tools they might have used produced just one small groove in the block I was working on. Not only that, I discovered that even a steel chisel was quickly blunted on basalt. Now obviously they had techniques of working with this stone of which we today are ignorant. Also, the block I was working on was of a particularly hard variety. Given more time and the opportunity to find a piece of stone that was slightly softer, it's possible the experiment could have been successful. But you know, Adrian,' he said, 'what to me is puzzling about Olmec art is not just the difficulty of working in a hard medium such as basalt.' He paused and looked at me with a conspiratorial glint in his eye. 'As I told them at the time, just carving a piece of basalt is not the most interesting question. The real difficulty, as I see it, is that archaeology has failed to find a sequence of development leading up to sculptures like the jaguar priests and Olmec heads. It's as if Michelangelo suddenly produced his David out of nowhere without a tradition of Greek and Roman sculpture to draw on. As a historian of the subject, I can tell you that sculpture doesn't work like that. The Romans copied the Greeks but even their sculpture didn't spring fully formed out of nowhere. It was preceded by that of Egypt, Mesopotamia and Asia Minor. The Greeks who carved the Parthenon frieze may have been the best in the world but they inherited a tradition that was thousands of years old.

'It's the same thing with the Olmecs. What I saw in Mexico was sensational from a sculptor's point of view. The execution was exquisite and exact. That doesn't happen by chance. There has to have been an evolutionary process of development leading up to this level of artistic sophistication. However, while you can readily see the influence of Olmec art on all that follows – on Aztec and

Mayan sculpture in particular – where are the antecedents of the Olmec heads? They are too good to have been the first sculptures of the Americas. Yet all the Olmec sculptures that have been found are of the same high standard. Why are we not finding much cruder examples belonging to an earlier period of this civilisation?'

These were ideas I had never heard expressed before. The fact that I was hearing them from the mouth of a professor of sculpture, a man greatly respected not just for his own work but for his knowledge of the history of the art, added importance to his comments.

At La Venta Park, besides the heads, there are a number of other sculptures. Among these are some examples of what are today referred to as 'altars', although they may have had some other function. These are not as tall as the heads but like them are carved out of single oblong blocks of basalt. These sculptures are, if anything, even more enigmatic than the ones in Professor Williams's pictures. I walked up to the first of these, which measured approximately 8 ft long by 4 ft wide and the same in height. Like others of its type, it seemed to represent a cave or grotto with a seated priest at the entrance. He was wearing a jaguar's head as a hat and sitting in a half-lotus position. I noticed that the jaguar motif was represented in many of the statues too, either as a headdress or carved separately as a protective deity. One block even features what you would have to call a were-jaguar: a curious sphinx-like creature that was part human and part jaguar.

The other beast to figure prominently in the art of the Olmecs was the rattlesnake. One of the altars, for example, had a gigantic serpent curled around it, its head at one end and its rattle at the other. That the rattlesnake was very important in the religion of the Maya is beyond doubt, but it was clear from these sculptures that reverence for these vipers was much older than the Classic Period.

On at least one of the altars there was a 'sky-band', a ribbon of glyphs representative of different stars and planets possibly symbolic of either the ecliptic or Milky Way. It was clear from this that they too were not a Mayan invention but went back to the dawn of Mesoamerican civilisation. It also indicated something else: that the Olmecs were keen astronomers. Further evidence of this was provided by another statue that was quite unlike all the rest as it represented either a monkey or possibly a monkey-man. Looking curiously uncomfortable, he was portrayed standing bolt upright and looking directly upwards at the zenith point. That this gave him some discomfort was indicated by the way he supported the back of his neck with his hands in a posture that most people adopt when studying the sky for any length of time. Seeing him, I nicknamed him the 'Monkey-astronomer', not quite realising at the time how apt this name might be.

Before leaving the park, I returned to take a second look at a group of statues – or rather reliefs – placed quite close to the entrance. I mentioned these in *The Mayan Prophecies*, suggesting that the first (known as 'the walker') could be a representation of Orion and that others, which seemed to involve warriors carrying children, perhaps recalled some great catastrophe. What I hadn't appreciated then, though maybe I should have, was just how similar certain, now rather worn, symbols carved on 'the walker' stone were to Mayan hieroglyphs. True, there are only three of them and they remain undeciphered, but they gave the lie to the idea that the Olmecs did not have a written language. Clearly they did. Our problem is that almost nothing of this language has survived. What has survived is a stele at Chiapa de Corzo in the Central Chiapas region of Mexico, which, though somewhat to the south of the Olmec heartlands of Tabasco, is generally regarded as epi-Olmec. It records a long-count date of 7 December 36 BC, which is the earliest long-count date yet discovered. This suggests that either the Maya learned of the long count from the Olmec or, more likely in my opinion, the 'Olmec' were themselves simply one branch of the Mayan peoples.

This would certainly be one explanation for the recorded fact that the native people living in the Tabasco region spoke a Mayan language at the time the conquistadores arrived. If the Olmec really were ancestors of the Maya, then it becomes more understandable why the latter adopted so many of their customs, such as pyramid building, playing the ball-game, the cults of the jaguar and rattlesnake, sky-bands and so on. Perhaps these were not so much 'borrowings' from the Olmec as common cultural characteristics that were shared by all the peoples of Central America from long before the rise of the Olmec civilisation at around 1100 BC. These elements of the Maya's cultural heritage seem to be homegrown, but others, for example knowledge of lime-based mortar, look very much like cultural imports. If this is the case, where did these cultural imports come from?

COMALCALCO, CITY OF BRICKS

In search of an answer to this question, my next port of call was the amazing Mayan city of Comalcalco. This city lies about 30 miles north-east of Villahermosa, a convenient distance for a day trip. As the coach drove along, we passed through several villages of wooden bungalows. Most of these had a wooden terrace to the front where people could sit under the shade of an extended roof. However, these terraces also had a more practical purpose: they were used for drying cocoa beans. As I had never seen these before, I asked the driver if we could stop and have a look. Rather than disturb total strangers, he said he would take us to visit a friend of his who ran his own chocolate factory.

This was not, as I feared, a large industrial complex but rather a hacienda. The owner turned out to be an affable man in his 60s. He explained how from scratch he had built up the business to the point where it was now a major force in the local economy. He invited us in and showed us around the factory, explaining how cocoa beans are turned into a bar of chocolate. Afterwards, as we sat on the balcony sipping at cold drinks, he brought out some bars of chocolate as well as its raw ingredient: cocoa pods. They were yellow in colour, about the size and shape of a small cantaloupe melon. Cutting one of these open, he showed me how it contained dozens of white beans, similar in size to an edible chestnut. 'Take one,' he said, offering me the opened pod. With some difficulty, I prised out one of the beans from its matrix of white, sticky pith. 'Don't bite it but just suck it,' he said, 'and tell me what you think.' Rather gingerly, I put the bean in my mouth, not at all sure what to expect. To my surprise, it was intensely sweet with a very delicate flavour similar to lychees. Smiling at the look on my face, he continued his lesson. 'This is one of the secrets of the cocoa bean,' he said. 'The old Mayans used to suck them when freshly picked to give an energy boost. These days, we soak fresh beans in water along with the pith. We ferment this liquid and then distil it to produce the spirit known as cacao. It is a valuable by-product and helps make the factory profitable.'

'What do you do with the beans themselves?' I asked.

'Round here, in the surrounding villages, many people have cocoa trees either in their gardens or growing in small fields. They harvest the beans and spread them out to dry on either the roofs or the porches of their houses. In the sunlight, the beans mature and turn brown. Some people keep them for their own use; others bring them here to the factory, where we turn them into chocolate. On a much larger scale, that's what the big manufacturers of chocolate do as well.'

'Is that what the old Mayans did?'

'Not exactly. The ancient Mayans used to grind the beans up and use them for making a drink. They didn't have sugar to sweeten it and they didn't have milk. Instead, they used to mix chocolate paste with ground-up chilli beans and water to make a beverage. It must have been quite hot and bitter to the taste, but that was how they liked it. They used to take jars of it out into the fields where they were working. It gave them energy and they were totally addicted to it. By the way, did you know they used cocoa beans as money?'

'Yes, I have heard that.'

'They not only traded beans among themselves but used to exchange them with the Aztecs, who were also addicted to the chocolate drink. It was the Aztecs who introduced it to the Spanish and they took cocoa beans back with them to Europe. They couldn't grow cocoa in Spain, the climate is not right, but they managed to keep a monopoly on the cocoa trade with America for a century before other people, like the English, managed to break it. As a matter of fact, your English word chocolate comes from the Aztec word for the beverage they drank: *chocolatl*.'

Leaving the chocolate factory, we got back into the coach and drove on to our real destination, the ancient city of Comalcalco. This turned out to be rather different from what

I was expecting, with a very modern visitor centre leading the way to the archaeological area, which lay in a verdant park. After walking about 100 yards, we arrived at our first pyramid, quite unlike any other I had seen before. Curiously, my immediate impression was that it looked like an ancient version of Battersea Power Station: a now disused 1930s building that as a boy I used to pass on the train when going into the centre of London. Today, like the pyramids of Comalcalco, Battersea Power Station stands abandoned, its huge shell empty. Nobody knows what to do with it but it can't be demolished because it is an icon of 1930s Art Nouveau styling. The pyramid reminded me of the former power station partly because of their similar bulk but mainly owing to the large expanse of windowless brickwork that make both stand out in their respective landscapes. I knew, of course, that what makes Comalcalco different from all other Mayan cities is that its pyramids and temples were made out of bricks. However, nothing had prepared me for what this meant in practice. Looking at this pyramid, as large if not larger than any in Palenque, gave me a strange feeling of familiarity. Though the city of Comalcalco is roughly contemporaneous and may even predate Palenque in foundation, the fact that it was made of bricks made it look much more recent in construction.

This feeling was amplified when I went over to another large building that is now called the Palace. This building, standing on top of a small eminence, contained rooms, brick pillars and even some small archways that seemed to serve a purely decorative purpose. The whole effect was highly reminiscent of Roman ruins I have seen in Britain at Richborough, Wroxeter and Wall. It is therefore not surprising that many people see Comalcalco as evidence for contact between the Americas and the Roman Empire.

There are indeed other reasons for thinking this besides surface appearance. First, not only were the pyramids and temples of Comalcalco built using bricks but these were themselves kiln-fired. Such kiln-fired bricks, which are very much stronger and more durable than sun-dried clay, were used throughout the Roman Empire. Even after the Western Empire collapsed, in the fifth century AD, such bricks continued to be made in Europe. Because bricks are so useful and durable, it is a technology that has lasted right up until our own times. While it can be argued that it is merely fortuitous circumstance that this technology was also developed in Mexico, it does seem that if this is the case, they were only used at Comalcalco. More curious still is that the one place where we find evidence of fired bricks being used was a city close to the coast with easy access by river. If the Romans were trading with the Maya, then this is just the sort of place to which they may have come.

This, it has to be said, is a highly contentious issue among Mexican archaeologists, many of whom, out of nationalistic pride if nothing else, are loath to admit cultural contacts of any sort between the Old and New worlds prior to the arrival of a few Vikings on the coast of Canada and the north Atlantic seaboard of the USA.

So is there any other evidence of Roman contact at Comalcalco? Well, yes. For one thing, not only were the bricks of similar size to those used by the Romans but they were also held together with a lime-based mortar. Now the Romans invented cement. They heated either limestone or chalk to high temperatures and then mixed the resulting lime with water and either sand or volcanic ash. This last ingredient mostly consists of a glassy compound containing silicon dioxide: the same compound that is more commonly found in sand but in finer particles. The Mayans of Comalcalco had much the same technology. They made their lime by roasting crushed oyster shells. Instead of volcanic ash, they did as we do today and simply mixed the resulting lime with sand and water. The formula, though not exactly like that used for Roman cement, makes use of the same basic chemistry.

As for the Comalcalco bricks themselves, they betray further evidence of cultural contact, for many of them were either stamped or inscribed prior to being fired. Close to the entrance of the archaeological site is a museum, small but modern and with some interesting exhibits. Among these were several samples of inscribed bricks that have been found during restoration of the pyramids and temples. As they were behind glass in a display cabinet, I was unable to examine them closely but they seemed to be about 18 in. to 2 ft long by around 10 in. wide and around 2 in. deep. In fact, they looked more like tiles. It is this similarity that gives the site its present-day name of Comalcalco, which means 'House of Plates' in the indigenous Nahuatl language of Mexico. I wrote about all of this in *The Mayan Prophecies* but it was quite something else to see them for myself.

In the early 1960s, a preliminary site-survey at Comalcalco revealed two bricks with inscriptions on them. When in 1976–8 Mexican archaeologist Pancio Salazar, working for the National Institute of Anthropology and History of Mexico (INAH), carried out more thorough excavations, some further 4,600 inscribed bricks were examined. Most of the inscriptions were recognisable as Mayan hieroglyphs but a few were different and excited some speculation. After Salazar's death in 1980, the collection was photographed and catalogued by archaeologist and epigrapher Neil Steede. He showed the pictures to Professor Barry Fell, then the leading light and founder of the Epigraphic Society. He was very excited by some of the bricks, writing in his seminal book *America BC* that one of them was a record of a Carthaginian moon calendar and that another was probably the work of a North African Berber as it carried an inscription that said 'Jesus Protector' in a Libyan script.

Fell also published a series of papers on the subject of the bricks for the Epigraphic Society's own journal: *ESOP*, or *The Epigraphic Society Occasional Papers*. In a paper entitled 'The Comalcalco Bricks: Part 1, the Roman Phase', which was published in Volume 19, he drew attention to what he calls masons' marks. Of the 4,600 bricks examined, some 1,500 were marked in this way. What was interesting about these marks was their similarity to Roman masons' marks found on similar bricks in Britain and other places. The Romans used such marks to keep a tally of individual productivity, each of their quarry slaves being required to produce some 200 bricks a day. By marking the bricks he made with his personal symbol, the slave could prove he was working at full speed.

The theory that Mexico was visited by mariners from the Roman Empire is not a new one. In 1933, a small sculpture of a head was found at Calixtlahuaca, about 60 kilometres west of Mexico City. What made this head special was that it is of a type easily identifiable as Roman of the period of Septimus Severus (c. AD 192–211). As it was discovered during an archaeological dig and was located well beneath two layers that anteceded the sacking of Calixtlahuaca by the Aztecs in 1510, its deposition had to predate the arrival of Cortés in Mexico in 1519. While it is in theory possible that it was brought to the Americas by the Vikings and somehow found its way to Calixtlahuaca through trade, it seems more likely that it was brought over by the Romans themselves.

Although the possibility that the peoples of the Valley of Mexico traded with Europe before Columbus is generally denied by Mexican archaeologists, there is much to commend the idea contained in the writings of the conquistadores themselves. Probably our best authority on what happened during Cortés's invasion of 1519 is a book written 20 years later by one of his soldiers: Bernal Diaz del Castillo. In the earlier part of his narrative, he writes of how on seeing a particular Spanish helmet an Indian ambassador of Montezuma claimed that their own ancestors had bequeathed them one that was similar:

> It happened that one of the soldiers had a helmet half gilt but somewhat rusty, and this Tendile noticed, for he was the more forward of the two ambassadors, and said that he wished to see it as it was like one that they possessed which had been left to them by their ancestors of the race from which they had sprung, and that it had been placed on the head of their god – Huichilobos [Huitzilopochtli], and that their prince Montezuma would like to see this helmet.[1]

Further on, Diaz tells us that this same ambassador, Tendile, reported back to Montezuma bringing with him an accurate drawing of Cortés. Montezuma was astounded to discover that Cortés bore a striking resemblance to one of his own chiefs:

> Tendile was the most active of the servants whom his master, Montezuma, had in his employ, and he went with all haste and narrated everything to his prince,

and showed him the pictures which had been painted and the present [the helmet] which Cortés had sent. When the great Montezuma gazed upon it, he was struck with admiration and received it on his part with satisfaction. When he examined the helmet and that which was on his Huichilobos, he felt convinced that we belonged to the race which, as his forefathers had foretold, would come to rule over the land . . . Then one morning, Tendile arrived with more than one hundred laden Indians, accompanied by a great Mexican Cacique [chief], who in his face, features and appearance bore a strong likeness to our Captain Cortés and the great Montezuma had sent him purposely, for it is said that when Tendile brought the portrait of Cortés, all the chiefs who were in Montezuma's company said that a great chief named Quintalbor looked exactly like Cortés and that was the name of the Cacique, who now arrived with Tendile; and he was so like Cortés, we called them in camp 'our Cortés' and 'the other Cortés'.[2]

Pictures of Hernán Cortés, including the lithographic frontispiece to my copy of Diaz's book, reveal him to have been typically European in appearance. His face was long, his nose aquiline, his eyes deep-set and his rather prominent chin sported a beard. He looks anything but Indian. It is therefore not out of the question that Quintalbor and maybe some other Aztec chieftains were descended, at least in part, from European or even Roman traders. Could they have been descendants of Romans who came to Mexico in the third century AD, bringing with them the Calixtlahuaca head and a helmet not dissimilar to the Spanish one sent to Montezuma by Cortés? We may never know the answer to that question, but it is certainly a possibility and one that on the evidence available we cannot discount.

The idea that the Romans traded with native Americans is further supported by the discovery of a number of hoards of coins in different places in the United States. Unfortunately, because these finds have been made accidentally and not under the controlled conditions of an archaeological dig, they tend to be ignored by the archaeological community.

This does not change the fact that such discoveries have been made and they tend to validate the idea that the Romans visited America. One such discovery was made in 1963 during building work on the banks of the River Ohio. A hoard of Roman coins was found by a contractor, who kept most of them to himself on the basis of finders-keepers. Fortunately, although he moved on taking both the coins and his identity with him he gave two of them to a fellow engineer. These passed to his widow, who later deposited them in the nearby Falls of the Ohio Museum in Clarksville, Indiana. There they were identified as bronze Roman coins: one from the reign of Claudius II (*c.* AD 260–8) and the other from the time of either Maximianus I (*c.* AD 235) or Maximianus II (*c.* AD 312).[3]

This find is interesting not just because of the coins themselves but on account of their location. The Ohio River is a tributary of the Mississippi, and that river disgorges into the Gulf of Mexico at New Orleans. It is recorded that the first Europeans settled in Clark County around 1797. These were pioneering folk and unlikely to have been collectors of Roman coins. The implication must be that some Roman or other European from after the year AD 260 brought the coins with him when he navigated his way up the Mississippi to the Ohio. The location of this find is also important as in the Ohio Valley there is a major concentration of artificial mounds thought by some to have been raised by post-Romano Britons.[4]

Professor Barry Fell provides further evidence of transatlantic contacts in Roman times. He records that Roman coins from the reigns of four consecutive emperors – Constantius II (AD 337–61), Valentinianus I (AD 364–75), Valens (AD 364–8) and Gratian (AD 367–83) – were found by a metal-detectorist within a space of one square yard on a beach at Beverly in Massachusetts.[5] Now it's always possible that the coins were either put on the beach by a hoaxer or lost by someone who happened to have such a treasure in his pocket while out walking the dog. However, both of these possibilities seem less likely than Fell's proposition that a Roman ship foundered off the coast in a storm and the coins were subsequently washed up on the shore.

Fell also writes about how in 1972 Punic (Carthaginian)

amphorae were found on the seabed off the Caribbean coast of Honduras and that in 1982 Roman amphorae (again dated to the third century AD) were discovered by divers off the coast of Brazil at Guanabara Bay, near present-day Rio de Janeiro. Amphorae are large jars that were used by the Romans and by the Carthaginians and Etruscans before them for transporting valuable goods such as wine, oil and salt. Many wrecks carrying such amphorae have been found in the Mediterranean but these were the first to come to light on the other side of the Atlantic. Unfortunately, it seems that, again for political reasons, the discovery of amphorae in the waters of either Honduras or Brazil is deemed unacceptable. We can but hope that before long, as the countries once under colonial rule but now independent democracies mellow with time, this attitude of cultural isolationism will change.

Fell found a great deal of evidence for cultural contacts across the Atlantic in even more remote times. According to Barry Fell and his colleagues, the so-called Micmac writing once used by the Wabanaki tribe of Maine is very clearly based on Egyptian hieroglyphs.[6] In the same book, *America BC*, Fell demonstrated how a memorial tablet called 'the Davenport Calendar Stele', which was found inside a burial mound in Iowa in 1874, is a sort of Rosetta Stone. The writing on it is in three scripts. He quotes an article by S.D. Peet that was published in 1892:

> The Davenport Calendar Stele was found in a burial mound in Iowa in 1874 by Reverend M. Gass, together with numerous other artefacts of North African and Iberian origin or relationship. This inscription is written in three languages, Egyptian hieroglyphs at the top, then Iberian-Punic from right to left along the upper arc, and Libyan from right to left along the lower arc. The Libyan and Iberian-Punic inscriptions say the same thing, namely that the upper hieroglyphs contain the secret of how to regulate the calendar. The hieroglyphs give this information by indicating that a ray of light falls upon the stone called the 'Watcher' at the moment of sunrise on New Year's Day, which is defined as the spring equinox in

March, when the Sun passes the first point of Aries. This stele, for long condemned as a meaningless forgery, is in fact one of the most important steles ever discovered, for it is the only one on which occurs a trilingual text in the Egyptian, Iberian-Punic, and Libyan languages. It is in the Putnam Museum, Davenport, Iowa, the repository of other priceless national treasures found by Gass.[7]

Champollion presented his decipherment of Egyptian hieroglyphs to the French Academy in 1822, so that by 1874 the subject was well established. Could it be possible that some genius linguist, living in Ohio in 1874, was able to master hieroglyphics well before 1874 and create and conceal a fragment of a stele with strange-looking scripts on

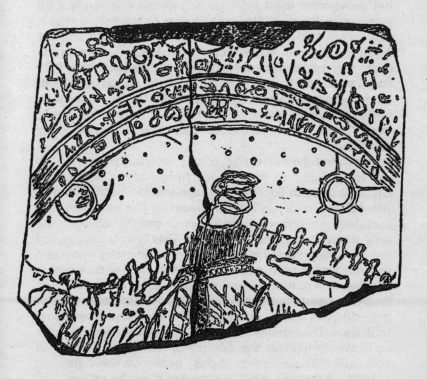

Figure 38: Davenport stele (after S.D. Peet, The Mound Builders, 1892)

it in demotic Egyptian, Iberian-Punic and Libyan? To anyone with an open mind, such a scenario seems even more unlikely than that the stele, a picture of which is reproduced in Fell's book, is genuine, and that it was buried in antiquity.

The only doubt must be whether or not the Egyptians were capable of sailing to the Americas. Yet this too is not as unlikely as it may seem. In 1970, the sceptics were given a bloody nose by the renowned Norwegian sailor and anthropologist Thor Heyerdahl. On only his second attempt, using a craft made of Egyptian papyri bound together in the traditional way, he and his crew sailed from the port of Safi in Morocco to the island of Barbados. The journey took them just 57 days.[8]

Of course, such a journey by modern-day navigators does not prove that the Egyptians or anyone else made such journeys in the past. However, it does at least show that it was possible. If Heyerdahl and his friends could sail the Atlantic in a boat made from papyrus reeds, then so too could the ancients. In fact, it would have been easier for them. Whereas Heyerdahl had to base the design of his boat on the wall paintings of Egyptian tombs, the Egyptians were well used to such boats. In building and sailing them, they would have known tricks of the trade that are today forgotten.

Nor would the ancient Egyptians necessarily have had to sail the Atlantic in boats made from papyri. Next to the Great Pyramid in Egypt is a strange-looking building that is possibly the most unlikely museum in the world. In it is contained a riverboat dating back to roughly 2500 BC: the time when the pyramid was built. This boat is constructed out of planks of wood rather than reeds, and they were sewn together rather than nailed. It could be argued that such a boat was not seaworthy and was only used as a river barge. However, parts of Bronze Age boats using a similar technology of sewing planks together have also been found in Britain. The remains of two of these were found in 1937 at North Ferriby on the estuary of the River Humber in Yorkshire. These boats have been carbon-dated to c. 1940–1680 BC. In 1960, a third boat was found that was even older than these, with a date range of 2030–1780 BC.

Though fairly flat, the boats had a sophisticated design with upraised prows and were capable of carrying up to 20 people. They were made from long planks sewn together with a 'thread' made from yew withies.

Now, while it could be argued that these too were riverboats, used for navigating up and down the River Humber, the same cannot be said for an even more recent discovery: a Bronze Age boat from Dover. This boat, found by archaeologists in 1992, was, like the others, made from planks of wood sewn together with yew withies. It was at least 14.5 metres long and 2.5 metres wide. Its estimated age is about 3,000 years, making it slightly younger than the Ferriby boats but still Bronze Age. More importantly, there is no doubt that it was seaworthy and used to carry goods across the Channel from France to Britain. Though such finds are few and far between, there can be little doubt that similar boats to this, a distant ancestor of the Viking ships, were used all over north-west Europe.

Though we have yet to find a surviving example, it is not beyond the bounds of possibility that the Bronze Age British also possessed larger ships, some of them capable of sailing across the Atlantic as well as the Channel. What we do know from ancient accounts is that the Phoenicians, the greatest seafarers of their age, regularly sailed around Spain to Cornwall. There they purchased the ingots of tin that were needed throughout the Mediterranean for the manufacture of the bronze which gave their age its name. To get to Cornwall from Spain means navigating the Bay of Biscay, which notoriously has some of the roughest seas in the world. If they could make this journey successfully and on repeated occasions, then we can be sure their ships were capable of sailing the Atlantic. Since their language was Punic (or even Iberian-Punic in the case of the Phoenician colony at Cadiz), there is no reason to discount the authenticity of the Davenport Calendar Stele simply on account of the remoteness of America from Spain. The Phoenicians had the ships, the navigation skills and the desire for trade that are the ingredients necessary for successful expeditions to Mexico. It is therefore not as improbable as it may at first seem that either they or their successors, the Carthaginians and Romans, visited

Comalcalco and instructed the local Mayans in the technology of brick-making.

Needless to say, Barry Fell's analysis of the Comalcalco inscriptions is just as controversial as his portrayal of the Davenport Calendar Stele as a Rosetta Stone. I was therefore curious to see if any of the bricks before me were the ones he described. As was perhaps to be expected, this proved not to be the case. Though there were some inscribed bricks on display, the pictures and writing on most of these bricks were clearly Mayan in inspiration. On one was depicted the image of a toothless old man who very probably represented the aged deity referred to by Mayanists as 'God L'. Another brick was embossed with an indecipherable inscription in Mayan hieroglyphs and a third showed the outline of a temple complete with roof-comb. More interesting and accessible than these was a very good likeness of a Maya nobleman of a type I had seen before at Palenque. He had his hair tied back to resemble the fronds emanating from an ear of maize and was wearing a large false nose. In fact, he looked very like Chan-Bahlum, the king of Palenque who built the Temple of the Cross. Beneath this portrait, which was of a head only, was a smaller version of another nobleman, shown in profile with one hand extended.

Leaving the obviously Mayan bricks, I looked to see if there were any tiles with Punic inscriptions of the sort mentioned in Barry Fell's book. If there were, I couldn't see them. However, there were several others that were of interest. One half of one was engraved with a picture of a cross made from two overlapping bones, which reminded me of the old skull-and-crossbones pirates' flag. Next to the crossed bones was a doodle of a snake that was in the process of unravelling. Beyond this was a zigzag pattern of alternating triangles. The triangles were filled with different designs, either of hatching or dots. The overall impression was reminiscent of the design on the skin of a diamond-backed rattler. Adjoining this was a pattern of alternating rectangles. Some of these were left blank while others were filled with varying numbers of dots. The meaning of this was not immediately apparent, but it reminded me of dominoes. Finally, at the edge of the tile was another series of triangles. These were smaller than the first and were left unfilled.

There was one last tile that drew my attention simply because of the intricate nature of the design, which involved curves and straight lines. However, it also resembled Arabic calligraphy when used as a decoration rather than straightforward text. So carefully had this tile been worked that it looked like it could be used as a printing-block for decorative wallpaper. This, however, was certainly not the case. More likely was that it was intended for use as a stamp to press repeatedly against wet stucco and thereby create a patterned band.

Taking the tiles as a whole, I was left with the impression that the pictures on them were mostly doodles. Unfortunately, none of the tiles that I saw was convincing proof that Comalcalco had been visited by sailors from either Europe or Africa – certainly not by the Romans. Then, just as I was about to leave the museum, something else caught my eye. There, in a cabinet, were several small sculptures of heads. What made them special was that they wore beards. Since full-blooded Native Americans don't grow facial hair, this in itself was suggestive of outside influences. However, the heads were also different from typically Olmec or Mayan art. Though they had slanting, oriental eyes, they had a very Caucasian look to them. One indeed looked Greek – very much like a sculpted mask that I had seen years ago in Greek ruins in Turkey. The Caucasian effect was further emphasised by the fact they were made of white stone. Strange as it may seem, another looked like an Afghan mujahid, complete with the sort of flat hat that is commonly worn in Afghanistan today. If this had been a museum in Kabul or the Panshir Valley, I would have assumed it was of local manufacture. This was not to suggest that Mexico had ever been visited in antiquity by people from Afghanistan but it was notable just how non-Mayan these heads were. They bore silent testimony in support of the account by Diaz concerning the helmet sent to Montezuma by Cortés, for they looked a lot more like Hernán Cortés than any Mayan Indian.

Seeing these heads and tiles, especially the one with a portrait that looked a lot like Chan-Bahlum of Palenque, reminded me of the Palenque foundation legend recorded by Father Ordoñez that a mariner called Votan from a place called Valum Chivim had made a number of journeys backwards and forwards across the Atlantic. It occurred to me that Valum comes from the Latin word *vallum*, meaning a palisaded rampart. Chivim would appear to be derived from the Latin word *civis*, meaning citizen. Valum Chivim must therefore mean a place of citizens with a wall around it. In the context of the Roman legacy, this would apply only to cities that enjoyed the status of Roman citizenship. One such city in Africa would have been Carthage.

The name Votan is a little harder to understand in a Roman context as it is clearly derived from Wotan or Woden, the German father of the gods. The connection of Votan with Africa at first seemed a little strange, until I remembered something rather important. In AD 406, a horde of barbarian tribes crossed the Rhine and invaded the then Roman province of Gaul. The leading tribe of this horde were the Vandals, who originated from the northern part of Germany and borderlands of Denmark. Deflected by a large army from Britain that came to the aid of the Gauls, the Vandal horde wept down into Spain. Some tribes, notably the Alans and Sueves, settled in Spain. The Vandals themselves went southwards, for a time settling in (V)Andalusia before crossing over the Straits of Gibraltar into North Africa, then a land of plenty. In 439, they seized Carthage and set about building an empire of their own. By 455, their empire, maritime in nature, was well established and they were able to sack Rome itself, plundering from it, among much else, the treasures of Jerusalem. However, the tide was soon to turn. In 532, Justinian, the Emperor of Byzantium, dispatched a large army to retake Africa. Because the Vandal fleet was away from port, the Byzantines were able to seize Carthage quite easily and lock the gates. The war rumbled on for a further 16 years, by which time the remaining Vandals gave up on Africa as a lost cause. A 'King of Africa' is recorded in British annals as leading a large contingent of Germanic pagans into Britain, but what happened to the rest of the Vandals is a mystery. Given this known history, it seems to me not at all unlikely that a Vandal leader called Votan should have led a maritime contingent from a 'walled city' – probably Carthage, though Leptis Magna or Tripoli would be other possibilities – across the Atlantic to America.

Barry Fell was of the opinion that Comalcalco probably preceded Palenque in its construction but that the two cities were closely related. If this is the case, then Father Ordoñez's Votan may have come to Palenque via Comalcalco. This is interesting, as in their book *A Forest of Kings* Linda Schele and David Freidel provide us with a genealogy of the kings of Palenque (see Figure 17). The first in this line is Bahlum-Kuk, the founder of the dynasty, and he acceded in AD 431. His great-great-grandson was Chan-Bahlum I, who died in AD 583. He had a daughter but not a son. Because of this, there was a change of dynasty around this time, his daughter's husband not being of the Bahlum-Kuk dynasty. This husband of Lady Kanal-Ikal, the great-grandmother of King Pacal the Great, remains unidentified. It seems to me that he is a definite candidate for being the mysterious Votan, perhaps a Vandal refugee from Carthage. This is not to dismiss the idea of earlier contacts between Africa and America in Roman times or earlier, but it might be an explanation for the Votan story as published by Father Ordoñez. However, the truth about the origins of Central American civilisation seems more complicated than journeys across the Atlantic. I remain convinced that it involves the lost civilisation of Atlantis, which Plato dated to before 9500 BC.

CHAPTER 9

The Aztecs and Atlantis

When I was writing *The Mayan Prophecies*, the subject of Atlantis was a recurrent theme. We explored the legend in its many variations and came to the conclusion that there really was an island civilisation that was submerged under the waves around the time Plato says. Before going into our reasons for saying this and its possible connection with the Mayan prophecy for the end of the age, it is worth reiterating the story from the start by beginning with our primary source for the legend: Plato.

The Greek philosopher Plato was a student of Socrates and wrote dialogues based on the latter's teachings. How much of his work can be ascribed to his mentor Socrates and how much was of his own invention is impossible to say. All we know for certain is that he revered the older man and used his teachings as a launch pad for his own writings. Atlantis appears in only two of Plato's works, the *Timaeus* and *Critias*. They appear to have been written at some time around 350 BC, towards the end of his life. The *Critias*, possibly the last dialogue he ever wrote, was a work in progress and remained unfinished at his death in 347 BC.

The *Timaeus* and *Critias* are mainly concerned with science. However, Plato introduces the scientific themes he wants to discuss by first telling the story of Atlantis. He bases his story on a tale apparently narrated to Socrates by Critias, Plato's maternal great-grandfather. Critias evidently heard it from his grandfather (another Critias) and he from his father Dropides. Dropides heard it first-hand from his good friend Solon, the famous lawgiver of Athens. Solon was born in *c.* 638 BC and died in *c.* 558 BC. Socrates died in 399 BC and Plato wrote the *Timaeus* in *c.* 350 BC. Thus, in telling the story of Atlantis, he was narrating a tale which he had heard fifth-hand and which was already nearly 200 years old when he set it down on paper. We should not be surprised, therefore, if the story contains some inaccuracies.

Before continuing this analysis, it is worth repeating what Plato actually writes about Atlantis. In the *Timaeus* – a reconstructed dialogue between Critias, Timaeus and Socrates – he tells us the following:

> 'There is (says he) a certain region of Egypt called Delta, about the summit of which the streams of the Nile are divided. In this place a government is established called Saitical; and the chief city of this region of the Delta is Sais, from which also King Amasis derived his origin. The city has a presiding deity, whose name is in the Egyptian tongue Neith, and in the Greek Athena, or Minerva. These men were friends of the Athenians, with whom they declared they were very familiar, through a certain bond of alliance. In this country Solon, on his arrival thither, was, as he himself relates, very honourably received. And upon his enquiring about ancient affairs of those priests who possessed a knowledge in such particulars superior to others, he perceived that neither himself nor any one of the Greeks (as he himself declared) had any knowledge of remote antiquity. Hence, when he once desired to excite them to the relation of ancient transactions, he for this purpose began to discourse about those most ancient events which formerly happened among us. I mean the traditions concerning the first Phoroneus and Niobe, and after the deluge of Deucalion and Pyrrha (as described by the mythologists), together with their posterity; at the same time paying a proper attention to the different

ages in which these events are said to have subsisted. But upon this one of those more ancient priests exclaimed, O Solon, Solon, you Greeks are always children, nor is there any such thing as an aged Grecian among you. But Solon, when he heard this, What (says he) is the motive of your exclamation? To whom the priest:—Because all your souls are juvenile; neither containing any ancient opinion derived from remote tradition, nor any discipline hoary from its existence in former periods of time. But the reason of this is the multitude and variety of destructions of the human race, which formerly have been and again will be: the greatest of these indeed arising from fire and water; but the lesser from ten thousand other contingencies. For the relation subsisting among you, that Phaeton the offspring of the Sun, on a certain time attempting to drive the chariot of his father, and not being able to keep the track observed by his parent, burnt up the natures belonging to the earth, and perished himself, blasted by thunder—is indeed considered as fabulous, yet is in reality true. For it expresses the mutation of the bodies revolving in the heavens about the earth; and indicates that, through long periods of time, a destruction of terrestrial natures ensues from the devastations of fire. Hence those who dwell on mountains, or in lofty and dry places, perish more abundantly than those who dwell near rivers, or on the borders of the sea. To us indeed the Nile is both salutary in other respects, and liberates us from the fear of such-like depredations . . . And from these causes the most ancient traditions are preserved in our country."[1]

The old priest goes on to tell Solon that at a time prior to the greatest of all floods, Athens was pre-eminent in war and was the best-governed city anywhere. He then tells the story of an invasion which occurred some 9,000 years earlier.

'"But though many and mighty deeds of your city are contained in our sacred writings, and are admired as they deserve yet there is one transaction which surpasses all of them in magnitude and virtue. For these writings relate what prodigious strength your city formerly tamed, when a mighty warlike power, rushing from the Atlantic sea, spread itself with hostile fury over all Europe and Asia. For at that time the Atlantic sea was navigable, and had an island before that mouth which is called by you the Pillars of Hercules. But this island was greater than both Libya and Asia together, and afforded an easy passage to other neighbouring islands, and it was likewise easy to pass from those islands to all the continent, which borders on this Atlantic sea. For the waters which are beheld within the mouth which we just now mentioned, have the form of a bay with a narrow entrance; but the mouth itself is a true sea. And lastly, the earth which surrounds it is in every respect a truly denominated continent. In this Atlantic island a combination of kings was formed, who with mighty and wonderful power subdued the whole island, together with many other islands and parts of the continent; and besides this subjected to their dominion all Lybia, as far as to Egypt; and Europe, as far as the Tyrrhene sea. And when they were collected in a powerful league, they endeavoured to enslave all our regions and yours, and besides this, all those places situated within the mouth of the Atlantic sea. Then it was, O Solon, that the power of your city was conspicuous to all men for its virtue and strength. For as its army surpassed all others both in magnanimity and military skill, so with respect to its contests, whether it was assisted by the rest of the Greeks, over whom it presided in warlike affairs, or whether it was deserted by them through the incursions of the enemies, and became situated in extreme danger, yet still it remained triumphant. In the mean time, those who were not yet enslaved it liberated from danger; and procured the most ample liberty for all those of us who dwell within the Pillars of Hercules. But in succeeding time prodigious earthquakes and deluges taking place, and bringing with them desolation in the space of one day and night, all that warlike race of Athenians was at once merged under the earth; and the Atlantic island itself, being absorbed in the sea, entirely disappeared. And hence the sea is at present innavigable, arising from the gradually impeding mud which the subsiding island produced." And this, O Socrates, is the sum of what the elder Critias repeated from the narration of Solon.'[2]

How Solon, the originator of Plato's story, came to hear about Atlantis is itself an interesting question. We know from various sources that he left Athens for ten years, during which time he visited Egypt. Saïs, the city where he is said to have met the old priest, was then Egypt's capital. It was a large city with a very long history going back to the pre-dynastic era. However, because it was mainly built from mud-bricks, very little remains of it today. This would not have been the case when Solon visited, for then it was at the height of its power and importance. Recent archaeological research reveals that the temple of Neïth was as large as the Temple of Amun at Karnac. So an old priest in charge of such a temple would have been a very important person. If anyone in Egypt was in the know about ancient traditions concerning an Atlantean invasion, it would have been him. King Amasis, who is mentioned in this account, was the last great pharaoh of Egyptian birth. He belongs to the 25th dynasty and ruled from 570–526 BC, which means he was pharaoh at the time of Solon's visit in c. 560 BC. Though this isn't mentioned in either the *Timaeus* or the *Critias*, the Greek historian Herodotus (c. 484–425 BC) writes that Amasis and Solon met. Given that Solon was probably the most famous Greek of his day and immensely respected as the lawgiver of Athens, that is not at all unlikely.

At the time of Solon's visit, Egypt was at peace with her neighbours and both men may not have been aware that a new power was about to rise in the east. Plato, who was writing 200 years later, had the benefit of hindsight. He knew that the reign of Amasis was a twilight period for ancient Egypt – a fact that added a certain poignancy to his tale. The Egypt of the later dynasties was no stranger to foreign invasions. It had suffered conquest by both the Assyrians and the Babylonians but had managed, with the help of Greek mercenaries, to break free. Under the 25th dynasty, Egypt was once more independent and had become a major trading nation. Here again the Greeks were central to the new prosperity. Granted a port of their own at a place near Saïs called Naucratis, they traded heavily with the Egyptians and were regarded as important allies. However, at the time of Solon's visit ominous clouds were already gathering on the horizon. Neither the priest he spoke to nor King Amasis himself may have realised it, but Egypt's days as an independent kingdom were numbered. In 526 BC, following the death of Amasis, Egypt was invaded by the Persians. Amasis's successor was killed in battle and Cambyses, the Persian king, declared himself pharaoh. Like a tidal wave, the Persian army had swept through Egypt and the ancient Grecophile states of Anatolia (Lydia, Caria and Ionia). However, the stubbornness of the Greeks – most especially the Athenians – dammed the flow. It was they who stopped the Persian Empire from engulfing the lands of the Mediterranean and would ultimately free Egypt from Persian oppression. It is with this background in mind that we have to read Plato's account of Atlantis.

It has to be remembered that for Plato, writing with the benefit of hindsight, the rise of Persia was recent history. He knew that it was the Greeks and most particularly the Athenians who had so far prevented the Persian Empire from extending into Europe. Athenian victories at the battles of Marathon (490 BC) and Salamis (480 BC) saw to that. However, the Persian threat had not yet gone away. The Persian Empire was still the mightiest the world had ever seen, stretching from the Aegean seaboard of Anatolia, through Mesopotamia and Iran to the borders of India. At the time Plato was writing, Egypt was at the end of a brief 'Late Period' (dynasties 28–30, 404–343 BC) in which it was for a short time independent once more of Persia. This would end in 343 BC when Artaxerxes III re-invaded Egypt and forced the country back into the Persian Empire. Ironically, it would be the Greeks, under Alexander the Great, who less than 20 years after Plato's death would drive the Persians out of Egypt and destroy their empire for good.

It was in the midst of all this history that Plato was writing. He was therefore not a disinterested chronicler of bygone ages but was someone keenly aware of the political situation unfolding around him. For him, the threat still posed by Persia to Greece and to Greek civilisation was all too real. Thus, in writing about Atlantis – a legendary adversary from times beyond memory – he was in a way presenting the idea of cyclical history. The Athenian victory over Atlantis could be looked upon as a forerunner of their more recent achievements in defeating the Persians at the battles of Marathon and Salamis. It is therefore clear that on one level his story of Atlantis was a parable for his own times: a prophecy of how the 'evil' empire of Persia would undoubtedly meet the same fate as Atlantis for daring to invade Greece and Egypt. But this is clearly not the whole story. For if the priest of Saïs was indeed telling the truth, and history was repeating itself, then this implied that the age in which Plato was living was probably coming to its end. This for Plato was a matter of more than a little importance, because the Greeks, like the Maya, believed that they were living in the last age of man.

THE LEGEND OF THE GOLDEN AGE

In our search for Atlantis, there are other, less obvious clues to Plato's intention in telling the story in the first place. When we examine his account of Solon's alleged conversation with the high-priest, the first question we have to ask ourselves is how did they get onto this subject in the first place? Plato himself furnishes us with the answer to this question when he tells us that:

> 'Hence, when he [Solon] once desired to excite them to the relation of ancient transactions, he for this purpose began to discourse about those most ancient events which formerly happened among us. I mean the traditions concerning the first Phoroneus and Niobe, and after the deluge of Deucalion and Pyrrha (as described by the mythologists), together with their posterity; at the same time paying a proper attention to the different ages in which these events are said to have subsisted.'

The discussion about Atlantis is therefore in the context of previous ages and world destructions. Solon must have been telling the Egyptians about Greek myths concerning earlier ages and the priest was responding with more detailed knowledge of his own. However, at the back of Plato's mind was doubtless the worrying idea that just as the defeat of Atlantis by Athens was followed by a catastrophe, so too the defeat of Persia (the modern 'Atlantis') by the Greeks of his own period would also be followed by a similar worldwide catastrophe.

For Plato, this may have seemed like a very real possibility. Strange as it may seem at first, the Mayan and Aztec idea that there were earlier ages and races of man than our own has its parallels in European mythology. These ages are described in a poem called *Works and Days* by Hesiod, a Greek poet who lived in the eighth century BC.

> First of all, the deathless gods who dwell on Olympus made a golden race of mortal men who lived in the time of Cronos [Saturn] when he was reigning in heaven. And they lived like gods without sorrow of heart, remote and free from toil and grief: miserable age rested not on them; but with legs and arms never failing they made merry with feasting beyond the reach of all evils. When they died, it was as though they were overcome with sleep, and they had all good things; for the fruitful earth unforced bore them fruit abundantly and without stint. They dwelt in ease and peace upon their lands with many good things, rich in flocks and loved by the blessed gods.
>
> But after Earth had covered this generation – they are called pure spirits dwelling on the Earth, and are kindly, delivering from harm, and guardians of mortal men; for they roam everywhere over the Earth, clothed in mist and keep watch on judgements and cruel deeds, givers of wealth; for this royal right also they received; – then they who dwell on Olympus made a second generation which was of silver and less noble by far. It was like the golden race neither in body nor in spirit. A child was brought up at his good mother's side a hundred years, an utter simpleton,

playing childishly in his own home. But when they were full grown and were come to the full measure of their prime, they lived only a little time in sorrow because of their foolishness, for they could not keep from sinning and from wronging one another, nor would they serve the immortals, nor sacrifice on the holy altars of the blessed ones as it is right for men to do wherever they dwell. Then Zeus [Jupiter] the son of Cronos was angry and put them away, because they would not give honour to the blessed gods who live on Olympus.

But when Earth had covered this generation also – they are called blessed spirits of the underworld by men, and, though they are of second order, yet honour attends them also – Zeus the Father made a third generation of mortal men, a brazen race, sprung from ash-trees; and it was in no way equal to the silver age, but was terrible and strong. They loved the lamentable works of Ares [Mars] and deeds of violence; they ate no bread, but were hard of heart like adamant, fearful men. Great was their strength and unconquerable the arms which grew from their shoulders on their strong limbs. Their armour was of bronze, and their houses of bronze, and of bronze were their implements: there was no black iron. These were destroyed by their own hands and passed to the dank house of chill Hades [Pluto], and left no name: terrible though they were, black Death seized them, and they left the bright light of the Sun.

But when Earth had covered this generation also, Zeus the son of Cronos made yet another, the fourth, upon the fruitful earth, which was nobler and more righteous, a god-like race of hero-men who are called demi-gods, the race before our own, throughout the boundless Earth. Grim war and dread battle destroyed a part of them, some in the land of Cadmus [Greece] at seven-gated Thebes when they fought for the flocks of Oedipus, and some when it had brought them in ships over the great sea gulf to Troy for rich-haired Helen's sake: there death's end enshrouded a part of them. But to the others Father Zeus the son of Cronos gave a living and an abode apart from men, and made them dwell at the ends of Earth. And they live untouched by sorrow in the islands of the blessed along the shore of deep swirling Ocean, happy heroes for whom the grain-giving earth bears honey-sweet fruit flourishing thrice a year, far from the deathless gods, and Cronos rules over them; for the father of men and gods released him from his bonds. And these last equally have honour and glory.

And again far-seeing Zeus made yet another generation, the fifth, of men who are upon the bounteous Earth.

Thereafter, would that I were not among the men of the fifth generation, but either had died before or been born afterwards. For now truly is a race of iron, and men never rest from labour and sorrow by day, and from perishing by night; and the gods shall lay sore trouble upon them. But, notwithstanding, even these shall have some good mingled with their evils. And Zeus will destroy this race of mortal men also when they come to have grey hair on the temples at their birth. The father will not agree with his children, nor the children with their father, nor guest with his host, nor comrade with comrade; nor will brother be dear to brother as aforetime. Men will dishonour their parents as they grow quickly old, and will carp at them, chiding them with bitter words, hard-hearted they, not knowing the fear of the gods. They will not repay their aged parents the cost of their nurture, for might shall be their right: and one man will sack another's city. There will be no favour for the man who keeps his oath or for the just or for the good; but rather men will praise the evil-doer and his violent dealing. Strength will be right and reverence will cease to be; and the wicked will hurt the worthy man, speaking false words against him, and will swear an oath upon them. Envy, foul-mouthed, delighting in evil, with scowling face, will go along with wretched men one and all. And then Aidos and Nemesis, with their sweet forms wrapped in white robes, will go from the wide-pathed Earth and forsake mankind to join the company of the deathless gods: and bitter sorrows will be left for mortal men, and there will be no help against evil.[3]

The idea of previous ages linked to gradually baser metals is echoed in the Book of Daniel. Roughly contemporaneous with Solon, Daniel could not have written his book before 586 BC (the date of the fall of Jerusalem to the Babylonians) but it may have been completed before 520, when the Jews rebuilt their temple. In his book, Daniel describes a symbolic statue which King Nebuchadnezzar saw in his dream:

> You saw, O king, and behold, a great image. This image, mighty and of exceeding brightness, stood before you, and its appearance was frightening. The head of the image was of fine gold, its breasts and arms of silver, its belly and thighs of bronze, its legs of iron, its feet partly of iron and partly of clay. [Daniel 2: 31–3]

Daniel goes on to explain the meaning of this vision and its relation to succeeding world empires.

> You, O king, the king of kings, to whom the God of heaven has given the kingdom, the power and the might, and the glory, and into whose hand he has given, wherever they dwell, the sons of men, the beasts of the field, and the birds of the air, making you rule over them all – you are the head of gold. After you shall arise another kingdom inferior to you, and yet a third kingdom of bronze, which shall rule over all the Earth. And there shall be a fourth kingdom, strong as iron, because iron breaks to pieces and shatters all things; and like iron which crushes, it shall break and crush all these. And as you saw the feet and toes partly of potter's clay and partly of iron, it shall be a divided kingdom; but some of the firmness of iron shall be in it, just as you saw iron mixed with miry clay. And as the toes of the feet were partly iron and partly clay, so the kingdom shall be partly strong and partly brittle. [Daniel 2: 36–42]

It seems doubtful that Daniel knew of Hesiod's poem directly, but it is likely that the idea of world ages and how these were symbolised by successively baser but harder metals was probably common currency throughout the Middle East and Europe during the sixth century AD. Whilst it is also unlikely that Plato read or even knew of the Book of Daniel, it is not impossible. Certainly, the idea of successive ages linked to progressively baser metals was not forgotten. The same idea is repeated some 500 years later by the Roman poet Ovid. In his *Metamorphoses*, written around AD 8, he mentions four rather than five ages, beginning with the Golden or Saturnian Age:

> The golden age was first; when Man yet new,
> No rule but uncorrupted reason knew:
> And, with a native bent, did good pursue.
> Unforc'd by punishment, un-aw'd by fear,
> His words were simple, and his soul sincere;
> Needless was written law, where none opprest:
> The law of Man was written in his breast:
> No suppliant crowds before the judge appear'd,
> No court erected yet, nor cause was heard:
> But all was safe, for conscience was their guard.
> The mountain-trees in distant prospect please,
> E're yet the pine descended to the seas:
> E're sails were spread, new oceans to explore:
> And happy mortals, unconcern'd for more,
> Confin'd their wishes to their native shore.
> No walls were yet; nor fence, nor mote, nor mound,
> Nor drum was heard, nor trumpet's angry sound:
> Nor swords were forg'd; but void of care and crime,
> The soft creation slept away their time.
> The teeming Earth, yet guiltless of the plough,
> And unprovok'd, did fruitful stores allow:
> Content with food, which Nature freely bred,
> On wildings and on strawberries they fed;
> Cornels and bramble-berries gave the rest,
> And falling acorns furnish'd out a feast.
> The flow'rs unsown, in fields and meadows reign'd:
> And Western winds immortal spring maintain'd.
> In following years, the bearded corn ensu'd

From Earth unask'd, nor was that Earth renew'd.
From veins of vallies, milk and nectar broke;
And honey sweating through the pores of oak.

But when good Saturn, banish'd from above,
Was driv'n to Hell, the world was under Jove.
Succeeding times a silver age behold,
Excelling brass, but more excell'd by gold.
Then summer, autumn, winter did appear:
And spring was but a season of the year.
The sun his annual course obliquely made,
Good days contracted, and enlarg'd the bad.
Then air with sultry heats began to glow;
The wings of winds were clogg'd with ice and snow;
And shivering mortals, into houses driv'n,
Sought shelter from th' inclemency of Heav'n.
Those houses, then, were caves, or homely sheds;
With twining oziers fenc'd; and moss their beds.
Then ploughs, for seed, the fruitful furrows broke,
And oxen labour'd first beneath the yoke.

To this came next in course, the brazen age:
A warlike offspring, prompt to bloody rage,
Not impious yet . . .

Hard steel succeeded then:
And stubborn as the metal, were the men.
Truth, modesty, and shame, the world forsook:
Fraud, avarice, and force, their places took.
Then sails were spread, to every wind that blew.
Raw were the sailors, and the depths were new:
Trees, rudely hollow'd, did the waves sustain;
E're ships in triumph plough'd the watry plain.
Then land-marks limited to each his right:
For all before was common as the light.
Nor was the ground alone requir'd to bear
Her annual income to the crooked share,
But greedy mortals, rummaging her store,
Digg'd from her entrails first the precious ore;
Which next to Hell, the prudent Gods had laid;
And that alluring ill, to sight display'd.

Thus cursed steel, and more accursed gold,
Gave mischief birth, and made that mischief bold:
And double death did wretched Man invade,
By steel assaulted, and by gold betray'd,
Now (brandish'd weapons glittering in their hands)
Mankind is broken loose from moral bands;
No rights of hospitality remain:
The guest, by him who harbour'd him, is slain,
The son-in-law pursues the father's life;
The wife her husband murders, he the wife.
The step-dame poyson for the son prepares;
The son inquires into his father's years.
Faith flies, and piety in exile mourns;
And justice, here opprest, to Heav'n returns.[4]

The inspiration for such later texts was undoubtedly Hesiod, whose works were studied nearly as avidly as Homer's account of the siege of Troy. What is more remarkable is the similarity between the Greco-Roman tradition concerning ages of man and what was recorded by the Maya in the *Popol Vuh*. Curiously, like one of the earlier attempts at making mankind that are described in the *Popol Vuh*, Hesiod's third creations (the men of the Bronze Age) are made from wood (ash trees). Though most of the details are different, the similarity in myths of earlier ages that have been terminated by the action of the gods is remarkable. The main difference between Hesiod and the *Popol Vuh* is that the former presents a steady degeneration of mankind, cycle on cycle, from the first race in the Golden Age to our present fifth race, whereas the latter portrays earlier races as much inferior to the present.

Hesiod's story of successive ages, beginning with a golden age of innocence, is consistent with other Greek myths. The most important of these concern the origins of the Olympian dynasty when Zeus (Jupiter or Jove) set himself up as king of gods and men. Prior to this time, the earth and heavens alike were ruled by his father Cronos (Saturn). According to the myth, Cronos was the son of Uranus, god of the sky, by the earth goddess Rhea. He was one of a number of 'Titans', an earlier race of gods who ruled the heavens and earth before the advent of the Olympians.

Cronos castrated his father and usurped his throne of heaven. According to Hesiod, Cronos married his sister Gaia. She bore him children but he, warned that like his father before him he would be overthrown by a son, ate them whole at birth. In this way, he consumed the gods Hades and Poseidon and the goddesses Hera, Hestia and Demeter.

Gaia grieved for her children and, aided by her mother, worked out a plan to stop Cronos from eating her last child, Zeus. We have already heard about this, but it is worth revisiting in more detail here. When Zeus was born, Gaia wrapped a large stone in swaddling clothes and gave this to Cronos in place of her baby son. Cronos swallowed the stone and Zeus was brought up secretly in a cave. When he reached adulthood, he determined to get his own back on his father. He persuaded the Titaness Metis to give Cronos an emetic. This induced vomiting and he threw up, whole and unharmed, Zeus's brothers and sisters. Cronos also brought up the stone, which thereafter was placed at the oracle of Delphi and referred to as the navel of the Earth.

Now a war for supremacy raged between the gods and the Titans. On the one side were Zeus, his brothers and some giants. On the other side were Cronos and most of his fellow Titans. The conflict lasted ten years, but in the end Zeus was victorious. After this, he and his brothers drew lots. Zeus (Jupiter) became king of the sky, Poseidon (Neptune) king of the oceans, and Hades (Pluto) king of the underworld. The above-ground earth was considered common territory which all three gods could visit as they pleased. Meanwhile, the fallen Titans were imprisoned in Tartarus, a hell-like gaol. Delegating some of his authority, Zeus set up a court of 12 ruling gods on Mount Olympus. Thereafter, these 'Olympians', whose number included Zeus's daughter Athena, who was born from his own skull, became the principal gods of the Greeks.

Though the Greeks and Romans revered Zeus as the father of the gods, their feelings for him were distinctly ambivalent. In their myths, they remembered that he was not the original creator of mankind. That honour belonged to Prometheus, one of four sons of a Titan named Iapetus. Prometheus, whose name means 'forethought', was clairvoyant and able to see into the future. He anticipated the defeat of the Titans and wisely put himself on the side of Zeus and the Olympians. However, when Zeus decided to destroy the human race, it was Prometheus who saved his progeny. He instructed Deucalion, the Greek Noah, to build an ark. Thanks to this, Deucalion and his wife survived the flood and in due course gave birth to a fresh generation of humans. Having failed to destroy mankind by water, Zeus now denied them fire in the hope that they would freeze to death. Again, Prometheus came to the rescue by bringing down fire from heaven and giving it to mankind. Zeus, still anxious to destroy the human race, now demanded that they offer him sacrifices of their food. He intended to starve them by taking from them the best parts of slaughtered animals. Once more, Prometheus came to the aid of his progeny. He arranged that two parcels of offerings be prepared: one for Zeus and the other for man. Zeus chose the better-looking offering, which turned out to be nothing more than offal and bones wrapped in skin. The humans were able to keep the choice cuts for themselves. Thereafter, much to Zeus's disgust, he was always offered the gristle and bones of sacrificial animals, while people were allowed to retain the meat for themselves. Prometheus also taught mankind many of the arts and sciences that they would need for life on Earth, including how to work with metals. However, he removed from them the gift of foresight or clairvoyance, which the earlier race of men had, as he knew it would only bring them sadness at their fate.

Enraged by the way Prometheus played tricks and continually disobeyed his commands, Zeus decided to punish him. He ordered that the Titan be chained to a rock on the top of a lofty mountain in the Caucasus. There an eagle came each day to eat Prometheus's living liver. As he was immortal, the liver grew back again at night ready to be eaten again the next day. This state of affairs might have continued for all eternity but fortunately for Prometheus he still had one trick left up his sleeve. Because he could see into the future, he was in possession of a piece of information that Zeus desperately needed: how to avoid the fate of his father and grandfather and not be overthrown by a son of his own. The only person who could help Zeus out of the sticky dilemma was Prometheus. He alone could look into the future and see what Zeus must do if he were to avoid castration and incarceration himself. Realising this, Zeus sent his half-human son Heracles (Hercules in Latin) to set the Titan free. Following Zeus's instructions, Hercules shot the eagle and removed Prometheus's chains.

In return for this act of kindness, Hercules was rewarded by Prometheus with some good advice. At the time he had received the call from Zeus, he had been engaged in one of the last of his so-called twelve labours. This task was that he should retrieve three golden apples from the Garden of the Hesperides and bring them back to the court of the King of Tiryns. The task was made more difficult because the tree on which the apples grew was guarded by a 100-headed serpent. Not only that but the islands of the Hesperides were beyond the edge of the world: far, far to the west, where the Sun travelled to after it set beyond the western fringes of Africa.

Getting the apples seemed like an impossible task. However, Prometheus, who could see all things, advised Hercules that there was one way. The gardens belonged to the daughters of his brother Atlas, another of the Titans. Unlike Prometheus, Atlas had taken part in the war against Zeus and the Olympians. As a result, he was condemned by Zeus forever to hold up one corner of the sky. Prometheus advised Hercules that he should visit Atlas and ask him to retrieve the apples for him.

Hercules did as instructed and, leaving Prometheus, made the long journey to north-west Africa. There, he cut a deal with Atlas that, if the Titan would agree to fetch the apples for him, he would temporarily take the burden of the sky onto his own shoulders. Atlas, who was delighted to be free of his responsibility even if only for a short time, agreed to the proposition. Leaving Hercules to hold up the sky, he crossed over the Atlantic and obtained the apples from his daughters. However, although he returned to Hercules as promised, he was unwilling to take back the sky. Hercules had to trick him into holding it up while he bent down to tie up his shoelaces. Then, with Atlas once more holding up the sky, Hercules made a quick escape, going back to Tiryns to deliver the apples to its amazed king. As for Atlas, he later looked unwittingly into the face of Medusa the Gorgon and was turned to stone. Today, transformed into the Atlas Mountains, he still supports the sky.

I have told these myths in some detail as I believe they offer fascinating clues to both the Greeks' ideas on astronomy and their own distant memories of Atlantis. Dealing with the astronomy first, it is evident that in this myth the Titan Atlas is to be equated with the constellation of Orion. As this constellation rises, it appears to push one half of the Milky Way upwards to eventually form an arc that passes through the zenith and spans the sky from south-east to north-west.

Then, as Orion sets in the west, so he appears to put down his burden of the Milky Way, which now forms a ring around the horizon. This, however, is not the end of the story, for now, rising in the east, is the constellation of Hercules. As he now climbs higher into the sky, so he appears to pull up the other half of the Milky Way to form a bright arc that this time runs from the south-west to the north-east.

Then, in a cycle that is repeated endlessly through time, Hercules sets in the west and Orion once again lifts his half

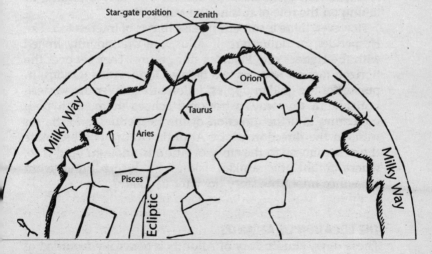

Figure 39: Orion supporting the Milky Way

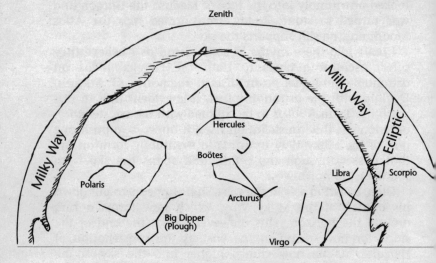

Figure 40: Hercules constellation pulling up the Milky Way

of the Milky Way into its elevated position. Astronomically speaking, then, it is true that as Atlas puts down his burden (the Milky Way) so it is taken up by Hercules. Both constellations symbolise sky-bearers and they alternate in taking on the role of raising the Milky Way.

However, there is something else that is of interest too. The Hesperides, or daughters of Atlas, are traditionally linked with the Hyades and Pleiades star-groups. They set over the horizon just an hour or two before Orion. Atlas (Orion), in pursuit of the golden apples, follows them over the western horizon. In this way, astronomy echoes myth, with both indicating that the direction of the Hesperides is in the far west, in the direction of the Atlantic Ocean. Is there more than astronomy to the story of Atlantis, though? And if so, where might the sunken island be located and what relevance might the story have for us today?

THE LOCATION OF ATLANTIS

These days, Plato's story of Atlantis is generally regarded as a myth without substance: an invention of his own that he created in order to describe ideas concerning the ideal city state. If the legend of Atlantis as a maritime empire is given any credence at all, it is considered to be a distant memory of the Minoan civilisation of Crete. This we know was destroyed around 1400 BC, when the neighbouring volcanic island of Thera exploded with such violence that it generated a massive tsunami. The argument then goes that this wave swept southwards, destroying the coastal cities of Crete. If the tsunami was anything like the size of the recent one in the Indian Ocean, it would have wreaked havoc throughout the Aegean, destroying coastal towns on islands and mainlands alike. This, it is claimed, is the truth behind Plato's story: a dim recollection of a geological event which, though local, was certainly catastrophic for both the Greeks and the island empire that held sway over Crete and against which Athens was at war.

The trouble with this, at first sight persuasive, theory is one of geography and timescales. No one disputes that Thera erupted and that the consequences were devastating to neighbouring islands. However, Plato specifically states that Atlantis lay beyond the Pillars of Hercules, which guard the Straits of Gibraltar. Beyond these straits was and is the Atlantic Ocean, so named after Atlantis. That Plato knew what he was talking about is amplified by his description that beyond the island of Atlantis itself lay another continent on the other side of the Atlantic Ocean. This can only be America, which implies that the Greeks of Plato's day, good mariners that they were, knew that it was there.

Until fairly recently, it was possible to imagine that Atlantis lay at the bottom of the sea somewhere in the mid-Atlantic. However, deep-water surveys reveal that there is no possibility of a sunken continent of the sort described by Plato in this region. On the contrary, there is the Mid-Atlantic Ridge, marking the boundary between two tectonic plates: the Eurasian and the North American. Although this ridge forms an underwater chain of volcanic mountains, these are nearly all far below the surface of the ocean. Only around Iceland does the ridge come up to the surface. Here, new land is forming – the converse of a sinking continent.

Even though the mid-Atlantic is unpromising as a location for Plato's Atlantis, there are tantalising clues which suggest that his story was not a complete fabrication but may have been based on memories of something real. A more specific clue to the location of Atlantis is what Plato had to say about its ruling dynasty. In his last dialogue, the *Critias*, he says that the island derived its name from its first king and that he was called Atlas. Now, as we have seen, Atlas was the father of the Hesperides and they lived on islands in the Atlantic in the direction of where the Sun set. According to Plato, the domain of Atlantis extended from America through these islands to Western Europe and North Africa. Could it be, then, that Atlantis was not really a single continent in the middle of the Atlantic but rather a maritime empire that spread across the Atlantic and embraced both islands and coastal countries on both sides of the ocean: countries and islands we now know as Mexico, Spain, Portugal, Libya, the West Indies, the Canaries, the Azores and maybe even the British Isles? If so, then how else might we learn about this antediluvian empire that once threatened Egypt and Greece? To find a possible answer to this question, I turned my attention to America's most famous twentieth-century psychic: Edgar Cayce.

ATLANTIS AND THE SLEEPING PROPHET

The quest for Atlantis has been the task – one might almost say obsession – of numerous scholars and enthusiasts. The influential book *Atlantis, the Antediluvian World*, written by US congressman Ignatius Donnelly in 1882, is a seminal work that has been drawn on by countless imitators. In the twentieth century, however, no one had a bigger influence than Edgar Cayce, the so-called 'Sleeping Prophet'.

The son of Kentucky farmers, Edgar Cayce was a sickly child who nearly died at age three when he fell on a nail and it pierced his cranium. He survived this trauma but throughout his life he suffered from a recurrent problem with his throat that caused him to lose his voice. His hoarseness seemed incurable until a family friend suggested they try hypnotherapy on him. He was put in a trance and to everyone's amazement it was found that in this hypnotised state he was able to converse quite clearly and tell them exactly what was wrong with him and how it could be cured. The evident success of this first attempt at hypnosis prompted further experiments with him. It was discovered that while in this state of trance he could answer questions of the most abstruse nature on subjects of which he had no previous knowledge, but when he woke up he could not remember a word of what had been said. People now began to request 'readings' from him. As all the readings he gave were carefully recorded by a stenographer, over time a vast archive of over 14,000 distinct reports was built up. At first, the readings were mainly concerned with finding cures for bodily ailments. However, it was soon discovered that, while in the hypnotised state, he could tell clients all about previous lives which they had lived and how those past experiences were influencing them in the present. Where the details of these previous incarnations could be checked, they were revealed as being accurate. However, many readings concerned lives lived so long ago that we have no written record of the times, let alone personal details of individuals living then. Included among these unverifiable records were numerous accounts concerning Atlantis.

It is now nearly 60 years since Cayce died, but in ways that he himself would have found surprising, his readings have become a major data resource. Of the 14,000-odd readings, only a few hundred mention Atlantis, but even so they make fascinating reading. What makes these readings useful and worthy of examination is the fact that, though they were given over a period of 20 years and Cayce was invariably asleep when they were given, they are internally consistent. Without going into the subject of Atlantis in any sort of methodical way, they nevertheless paint a picture that is consistent both with Plato's account and also with geological reality. Cayce's Atlantis comes to its watery end around 10,500 BC, i.e. about 1,000 years before Plato's end-date of *c.* 9500. However, according to the sleeping Cayce, this aquatic finale was only the last of a series of catastrophes to hit the island. The readings claim – and this is, of course, impossible to verify – that mankind came upon the Earth roughly 10.5 million years ago. Like Plato, Cayce claimed that the Atlantean civilisation became corrupt, as people put materialism before the service of God or 'the gods'. This corruption was punished by a series of upheavals, beginning around 43,000 BC. A second and greater upheaval took place around 28,000 BC. This split up what had been a single continent into a collection of islands. The largest and most important of these he called Poseidia. It was in the Bahamas – the Bimini Islands off the coast of Florida being then mountain peaks on Poseidia. The capital city of Atlantis was also on Poseidia, presumably on the then coast. It sank under the waves in the final phase of destruction, which, according to Cayce, took place around 10,500–10,000 BC.

Though some of what Edgar Cayce had to say about Atlantis and the Atlanteans seems more like science fiction than fact – e.g. that they were in possession of technology more advanced than our own, with flying machines, electricity and even a death ray – his placing of the island of Poseidia in the Bahamas rings true geologically. For whereas the middle of the Atlantic Ocean is characterised by some of the deepest water in the world, the sea in the area of the Bahamas – the Bahama Banks – is very shallow. This agrees with Plato's statement that when Atlantis went under the waves, the sea in that area became unnavigable on account of mud just below the surface.

In his trance state, Cayce claimed that prior to the destruction of Atlantis, some of the Atlanteans escaped. Anxious that the memory of their homeland should not be entirely forgotten by future generations, they buried records of their civilisation in three places. One set of records was secreted in Egypt, one in the Yucatán Peninsula of Mexico and one in Atlantis itself.

To date, the Egyptian 'Hall of Records' is the one which has attracted the most attention. Several missions have been launched in search of it, mostly mounted under the auspices of the Association for Research and Enlightenment (A.R.E.), the organisation founded to perpetuate Cayce's work. The general consensus is that there is a secret chamber, as Cayce said, in the vicinity of the Great Sphinx of Giza. Ground-penetrating radar surveys have revealed below-surface anomalies that are consistent with such a chamber. However, to date the Egyptian authorities, though at times cautiously supportive of this work, have been unwilling to allow actual digging to commence. One reason for this

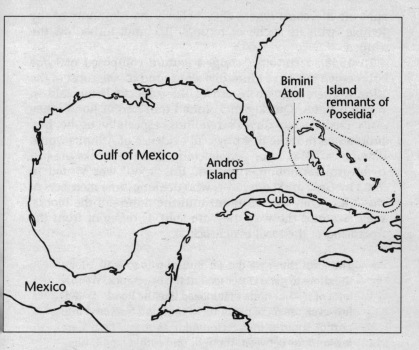

Figure 41: The location of Atlantis in the Bahamas area

caution has been the political situation both within Egypt itself and throughout the wider Middle East region. In the current climate of terrorist hostility towards all things Western, it has been deemed unwise for the Egyptian Antiquities Organization to allow the work to go further. For the time being, the search for Cayce's elusive chamber of records under the Sphinx has been suspended.

One result of this impasse in Egypt has been to refocus attention on Atlantis itself and, more specifically, the location of the cache of records in Bimini. Cayce himself visited Bimini on more than one occasion and while there gave several trance readings that reaffirmed this as a portion of Atlantis. He also said that in its vicinity was to be found an ancient pyramid-temple containing the Atlantean cache of records. Nobody has yet been able to find this temple, though other people (notably George McMullen, a Canadian psychic and psychometrist) have pinpointed a spot off the coast where they believe the remnants of the temple, with its cache of records, lies undisturbed on the seabed.

In 1968, a curious J-shaped feature composed of large submerged blocks of limestone was found off the coast of the island. There was no record of these blocks being laid in modern times. Quickly nicknamed the 'Bimini Road', they gave credence to Cayce's readings, especially as he had prophesied that the first physical evidence of Atlantis would emerge in 1968. Some people claim that the blocks are just a natural formation. However, this 'road' was visited in 2003 by Greg and Lora Little, who are long-time members of the A.R.E. They are adamant that the nature of the blocks, their size and the way they are split, is different from the fracturing of the local beachrock.

> The final thing we did on Bimini was to walk along the shore to get a closer look at the beachrock. We saw a lot of it – but none of it looked like the Road. We did, however, see a couple of rocks that had fractured into partial squares and rectangular shapes. These were few and far between, though, and rather small, none larger than 6 or 7 ft. We knew that beachrock can fracture into near-perfect squares, with right angles, so seeing these couple of square rocks wasn't a surprise. What was a surprise was that there were so few of them and that they were thin, much less than a foot thick. Publications which assert that the Bimini Road is composed of naturally occurring fractured beachrock point out that there is a lot of beachrock on the island that is *identical* to the Road. These articles give the impression that there are long rows of square and rectangular beachrock lying on the surface at Bimini. This is just not true. The composition of some of the stones on the Bimini Road is identical to some of the beachrock on Bimini, but there aren't many square or rectangular blocks of 2-ft-thick beachrock to be seen.[5]

Since 1968, the A.R.E. has been involved in numerous attempts to find more evidence for Atlantis. In 2003, and again in 2005, the Littles made a systematic search of the Bahamas. They found further compelling evidence for a lost civilisation when exploring the island of Andros.[6] This island, the largest of the Bahamas, lies to the south of Bimini. Largely uninhabited and uninhabitable, it is the location of the remains of what was clearly a very ancient temple on what was very likely the highest point of the island. The Littles also discovered that one of the main reasons why the west coast of the island remains almost uninhabited is the shallowness of the water. Even some miles offshore, it is still only waist-deep and the sea-bottom is covered by a thick layer of slimy mud. As the interior is covered with mangrove swamps, the west coast is cut off from access except by very shallow-draught boats.

This is all very interesting. A topographical map of the Bahamas reveals that at the time of the last ice age Andros and Bimini would have been small parts of a much larger island. This island, presumably 'Poseidia', would have embraced most of the Bahamas. It would have looked rather like a letter 'S' turned on its side with two large inlets of water. Nassau, the current capital of the Bahamas, is positioned at the entrance of one of these inlets. Bimini would have been at the tip of the 'S'. Andros Island would have been on high ground on the largest part of the island, which then would have extended westwards almost as far as Cuba.

Cuba itself, which other researchers, such as Andrew Collins, identify with Atlantis, would also have been much larger than it is today – as would Florida and the Yucatán Peninsula of Mexico. The Central American bulge around Honduras and Nicaragua would extend far out to sea, more than halfway to Jamaica. Since we know that the ice age ended around 11,000 BC, the evidence would seem to be compelling that if there is any truth in the story of Atlantis sinking beneath the waves, it must be associated with this change of sea-levels in the region of the Bahamas and West Indies.

It is easy to see why early explorers in the Mayan lands, people like Brasseur de Bourbourg, equated the ruins of their abandoned cities with the lost civilisation of Atlantis. The idea that mariners may have visited America thousands of years before the epic voyage of Columbus in 1492 also inevitably brings up the question of the lost continent of Atlantis. For if the Carthaginians or Romans could have sailed across the Atlantic from Africa, why should not other mariners have crossed the other way? In short, could there once have been an 'Atlantean' empire, older than the Greek or Roman, that had its centre in the islands of the West Indies?

In Brasseur's time, the Mayan hieroglyphs were untranslated and seemingly untranslatable. The chronology of civilisations in the Americas was unknown and so too was the identity of the builders who left behind such enormous monuments as the pyramids of the Sun and Moon at Teotihuacan. That the Atlantic was deep was not in dispute, but prior to the development of the theory of plate tectonics it seemed not unreasonable to suppose that beneath those inky depths there might lie a sunken continent. Most of these suppositions we now know to be wrong. Carbon dates from Central America reveal that the Mayan civilisation was barely in its infancy when Plato was writing his dialogues. Most Mesoamerican cities, places like Palenque and Chichen Itza, were not even founded when he was writing. Thus, to attribute their cities to the Atlanteans is just as false as crediting them with building ancient Rome. To find the truth, we have to separate geography from history, actuality from wishful thinking. If we can do this and yet still find ourselves left with a residue of unexplained facts that fit well with the myth, then and only then can we begin to be convinced that there is some truth behind the myth of Atlantis.

The association of the legend of Heracles and the golden apples with the Bahamas and Antilles makes sense of the Mayan and Aztec myths concerning earlier ages. In the Hercules story, the islands of the Hesperides are described as being in the farthest west, where the Sun goes to during the night. We are also told that these hard-to-get-to islands were a sort of paradise on Earth. Exotic fruit trees grew there and the climate was so pleasant that three crops per year could be raised. That seems like a perfect description of the West Indies, where fruits of all sorts grow in abundance. There are further associations too. In the Greek myths, Atlas is not only a Titan who holds up the sky but is also named as the first king of Atlantis. If that is the case, then it follows that his daughters, the Hesperides, lived in Atlantis – or rather on islands that could be regarded as remnants of Atlantis after the great disaster had engulfed most of the lost continent. The Canary Islands and Azores are often put forward as candidates for their island home. However, I do not believe that either of these island groups, which are actually volcanic peaks, are what Plato had in mind as his 'Atlantis'. In my view, a more probable location was the West Indies, where, as he says, navigation is hazardous because of shallow waters. Here, apart from during the hurricane season, the climate is indeed delightful. Not only that but tropical fruits grow in abundance on trees that cannot be grown in the area of the Mediterranean. The 'golden apples' which Hercules had to retrieve could be referring to nothing more magical than the humble coconut. This would have been an unattainable delicacy as far as the ancient Greeks were concerned. Alternatively, they may have been cocoa pods. As I discovered on my visit to the chocolate factory near Comalcalco, these are golden in colour when first picked and contain a sweet-tasting pith. The beans they yield were used to make a beverage that was highly prized by both the Aztecs and the Maya. Most people know that, like coffee, cocoa is addictive as a stimulant. It is also said to have aphrodisiac properties. If this had been known to the Greeks, then it is understandable that the 'golden apples' should originally have been a wedding present to the goddess Hera from her husband Zeus. Perhaps he was intending that she should have a nightcap of hot chocolate to put her in the mood. Jokes aside, the story of Hercules' quest for the golden apples of the Hesperides could have been based on a distant memory of a time when the people of the Mediterranean traded with the New World for cocoa beans.

This, I believe, might make a rewarding line of research for some intrepid investigator: when were the first cocoa beans brought back to Europe? At any rate, Plato's report about Atlantis seems to record a vague memory by the Egyptians if not the Greeks themselves that there were tropical islands in the farthest west. Associated with the name of Atlas the Titan, exotic fruits grew on these islands. Given this association, it seems not unreasonable to believe that the so-called war of the gods, when the Titans rebelled against Zeus's coup against his father Uranus, may refer to a much lesser event: the invasion of Europe and Africa by the people of Atlantis. If this occurred around 10,500 BC, then the destruction of Atlantis can be explained as being associated with the rise in sea-levels around that time: the ending of the last ice age. If the ice melted over a very short period, then such a catastrophe could have occurred quickly and without much warning. However, even if the melting was slow and for a time the Atlanteans were able to hold back the sea with levees, they could not have prevented the inevitable forever. We have recently seen something of the power of tropical storms in the region of the Gulf of Mexico with the flooding of New Orleans. A Bahamas-based city of Atlantis would have been just as vulnerable if all it relied upon for protection from the waves was a system of dykes.

The one advantage of a 'slow' melt as opposed to an immediate catastrophe (e.g. a massive earthquake, a collision with an asteroid or the explosion of some sort of nuclear power plant) is that it would have allowed time for preparation. This fits the Cayce scenario and gives substance to his claim that the Atlanteans, knowing that their culture was doomed, deliberately placed records of their civilisation in Mexico and Egypt as well as in Bimini. It is to be hoped that we will one day find these records, but in the meantime we can look at other evidence of the Atlanteans: most specifically traces of their influence in the customs and traditions of the Maya. This was the next step on my quest.

CHAPTER 10

The Serpent Cult and the Planetary Gods of the Maya

It was good to be back in Merida, the city founded by Don Francisco Montejo and today the capital of Mexico's Yucatán Province. Once more, I checked into the Casa Balam Hotel, a baroque establishment close to the cathedral and at the very heart of the Old City. This hotel did not exist at the time Stephens and Catherwood visited the city but they would nevertheless have felt at home staying there. Although built, perhaps, in the 1960s, it gave the appearance of a typically colonial '*casa*'. At its centre was an atrium with a small fountain playing on some rocks. Potted plants and baroque chairs arranged in groups around coffee tables completed the illusion of nineteenth-century colonial living. Unlike the bedrooms, which faced onto the noisy street below, the atrium was an oasis of calm.

On enquiring, I was told by our guide that the hotel came by its name from its founder: a legend in the neighbourhood who was known by the nickname of Balam, or jaguar. Although he was a major figure in the early days of Yucatán tourism, there was, alas, no connection with Chan-Bahlum, the famous king of Palenque who built the Temple of the Cross. My immediate interest, however, was not the hotel or even the city but to contact a remarkable gentleman whom

I had met the last time I was in Merida. A couple of phone calls and a taxi ride saw me outside his little house, which lay in the German quarter.

Sitting outside enjoying the cool of the evening was an elderly man wearing an open-necked shirt and loose-fitting trousers. Although he looked older and more infirm than I remembered from my last visit some four years earlier, he was in remarkably good form for a 91 year old. 'Good evening, Mister Gilbert,' he said, shaking me by the hand.

'Good evening, Don José,' I replied and turned to introduce him to two young men I had brought with me from the group. 'This is Don José Diaz Bolio, pioneer of the understanding of the role of the rattlesnake in Mayan art.'

They too shook his hand and he led us into his little parlour, which doubled up as a dining room. On the table were spread out a number of his books and pamphlets with titles such as *The Geometry of the Maya and their Rattlesnake Art* and *Why the Rattlesnake in Mayan Civilization*. He beckoned us to be seated. 'Let me first say how good it is to see you again, Mister Gilbert,' he said. 'I want to thank you so much for what you wrote in your book *The Mayan Prophecies*. As you see, I have it here.' He held up a copy of the book: one I had sent over a couple of years earlier with a London artist who was keen on meeting him to receive a personal initiation into the mysteries of rattlesnake art. 'I cannot speak for the rest of your book,' he continued, 'but the chapter you wrote about my work is exactly correct. Thank you for spreading this knowledge to the world in a way I have not been able to do myself.'

I felt self-conscious at these words of undeserved praise but he was effusive in his thanks for my having helped him to circumvent his critics in Mexico and give his ideas an international airing. 'You know,' he said, turning to the two young men I had brought with me, 'I have had no end of opposition from the authorities in Mexico.'

'Why's that?' one of them replied. 'Surely people must be pleased that you have made important discoveries.'

'That's what I thought when I started out, about 50 years ago now. I thought scientists were meant to be unbiased in the way they assess evidence but they find my work insulting to Mexico. The reason is that I discovered that at the heart of Indian culture stands a giant rattler. You will find it in their art from the tip of Tierro del Fuego to northern Alaska. The rattlesnake, especially the species *Crotalus Durissus Durissus* – by the way, the only one native to the Yucatán – was the inspiration behind the entire Mayan civilisation. Before I remarked on it, nobody seems to have noticed that the entire culture and art of the Maya was based upon close study of this serpent.'

'How did you make such an amazing discovery?'

'Me? It all started when I visited the ruins of old Chichen Itza in 1942. I saw a temple-lintel with rattlesnakes carved on it. After that, I started seeing rattlesnakes everywhere in the art of the Maya. I spent three years then, searching to see if anyone else had noticed the importance of the rattlesnake in indigenous art. It soon became clear that from the time of the Spanish conquest to now nobody had.'

'No kidding. What happened then?'

'In 1945, I met a famous anthropologist, Professor Enrique Juan Palacios. He was a teacher of archaeology at the Instituto Nacional de Antropologia e Historia. I showed him the results of my researches, including some pictures. For some minutes, he stared at the pictures in amazement. He congratulated me on my discovery and then said, prophetically as it turned out, that although I had found what they had all been seeking for a hundred years, nobody would admit I was right. It would take two generations before my ideas would gain acceptance. I wasn't that patient, so in 1955, after much further research and exploration, I published my major work: *La Serpiente Emplumade: Eje de Cultures.*[1] The book was filled with pictures and diagrams to prove my point. It presented incontrovertible evidence that the religion of the Maya and all other pre-Columbian civilisations was at heart a cult of the rattlesnake.'

'I don't understand,' said my friend. 'If you had discovered what they had all been seeking for a hundred years, surely they would be pleased?'

José laughed. 'You have to realise that what I was proposing was tantamount to heresy – not so much against Christianity as against the new religion of archaeology. At that time, the archaeologists and anthropologists had an idealised vision of the Maya. Unlike the bloodthirsty Aztecs, who are known to have practised human sacrifice, the ancient Maya – those who left behind the romantic ruins – were looked upon as philosopher-priests of the type described by Plato. The scholars did not like to hear that their pacifistic liberal Maya worshipped the Sun in the form of a great rattlesnake or that they too fed this god with the *vis vitae* of human hearts. They thought my discoveries were an abomination and publishing them was considered disrespectful to the Mexican people. So they did their best to suppress these ideas. As no Mexican publisher would handle my books without the approval of the academic authorities, I had no choice but to print them myself. Unfortunately, I was a widower with a large family to support and very little money. But I didn't let that stop me. I started a business of my own: extracting essential oils from plants. It just about provided us with a living and left me with a little money to pursue my passion. I was able to start my own publishing business, which I called "Area Maya". Since nobody else would handle them, I printed the books myself.'

'What, you mean you actually printed them by hand?'

'Yes. I had very little help. My only mistake was that I wrote my major works in Spanish. I should have written them in English. They would have reached a much wider audience if I had. I didn't make the same mistake when I published my small introductory books. These continue to sell well here in Merida and even in some other places.' He gestured towards several piles of booklets with titles such as *The Mayan Calendar Made Easy* and *The Geometry of the Maya and their Rattlesnake Art*.

'Yes, we read about these in *The Mayan Prophecies* and saw a couple of copies in a bookshop close by our hotel just before coming out,' said my other friend. 'Can you tell us a bit more about your discoveries?'

'That would take all night,' continued Don José. 'You will need to read my books, especially *La Serpiente Emplumade: Eje de Cultures*, to get a full understanding. However, put briefly, you could say that the Maya, and other cultures influenced by them, held the rattler Crotalus Durissus Durissus in highest esteem because it personified so much for them that was culturally significant. They called it *Ahau*

Can, which means Lordly Serpent, and they studied its ways in intimate detail. For example, they noticed that it changes its teeth every 20 days. This gave them the time period they called the Uinal. The glyph for a Uinal shows an open serpent's mouth with two prominent fangs. They also thought of the Sun as being a gigantic plumed serpent that coils itself around the Earth. Each day it rises in a slightly different place on the horizon, moving first towards the north and then back to the south. The pattern of the Sun's movements was for them similar to those of a serpent coiled around the Earth. On two days in the year, it would be seen to cross the zenith point. On those days, the stone serpent heads that stand out horizontally on the ball-court at Chichen Itza cast a shadow on the ground that is exactly the same size as the heads. I know this because I have seen it myself, but that is not all . . .'

'Excuse my ignorance,' interrupted my first friend, 'but when would that happen?'

'At Chichen the first day of zenith transit is 22 May and the second is 16 July. It was the second one that was most important to the old Maya, as around this time their sacred serpents shed their skins. They are reborn in a new skin just like the Sun is reborn on its zenith transit. Each time they do this, they grow an extra rattle. These rattles are shaped like a human heart. The Maya still keep these rattles if they find them, by the way. They regard them as good-luck charms, rather like Europeans hang up horseshoes. For all these reasons, they made the second and not the first zenith transit the start of their new year.'

'Have you ever seen a snake moulting its skin?'

'Many times. Years ago, I used to keep rattlers so that I could study them.'

'What, here?' My friend looked alarmed.

'In this very house, but don't worry, I don't have any snakes these days. I gave up with them after one escaped and frightened my neighbours.' Again, he laughed. 'Keeping rattlers as pets, I was able to observe them very closely. As I said, when the snake moults its skin, which in the wild it does only once a year, it grows an extra rattle as well. So each individual rattle represents a year in the life of a Crotalus serpent. The Mayans count the number of rattles to calculate the age of an individual rattler like we count tree-rings to find the age of a tree.'

'That's amazing,' said my other friend. 'Perhaps that's why they put rattlesnakes on the Aztec calendar stone.'

'I'm glad you mentioned that,' Don José continued. 'There is some confusion about that stone. What are described in the guidebooks as drops of human blood are in fact stylised rattles. There are fifty of them and if you add two more, for the twinned serpents round the outside, it comes to fifty-two.'

'That's the number of years in a sheaf, isn't it?'

'Exactly so. The Aztecs, like the Maya, used to divide time into 52-year periods. They believed that the end of the world will come at the conclusion of one of these 52-year cycles. That's why every 52 years, they would go out into the mountains and anxiously watch the movement of the Cabrillas.'

'What are the Cabrillas?'

'The Pleiades stars. The Spanish call them Cabrillas, which means little goats. Don't ask me why they call them that because I don't know. The Maya watched these stars too. They called them the *tzab*, which means 'snakes-rattle'. But as I was saying, the Aztecs watched these stars intently on the last night of a 52-year period. If at midnight they carried on moving past the zenith, then they knew that the world was not going to end and they had at least another 52 years to go. They would celebrate by lighting fires and making human sacrifices.'

With that, he stopped speaking and got up. He walked over to a cupboard and brought out from it a polythene bag. 'I have something here to show you all that you may find interesting.' He opened the bag and drew from it what at first looked like two long strips of leather. 'These are rattlesnake skins from the genus Crotalus Durissus Durissus,' he continued. 'The most important cultural influence of the Crotalus is not its rattle but the design on its skin. As you can see, they have a very definite pattern of markings on their backs. If you analyse the pattern carefully, as I have done, you will discover that the square pattern is repeated over and over again. It is based on a shape I call by the name *Canamayte*.'

'What does Canamayte mean?' said my second friend.

Figure 42: Canamayte pattern

'It's Mayan for snake-square,' the old man replied. 'Put simply, it's a square with a cross dividing its sides in such a way that the square contains four smaller squares. This square with cross, the Canamayte, is the most important design motif in Mayan art. You will see it repeated

Figure 43: Canamayte pattern of snakeskin compared with pyramid detail

everywhere with many variations. Even today, it is the geometric shape underlying the design of the embroidery on the traditional Mayan women's dress called the huipil.'

'What size is the square on the snake's back?' I asked.

'That depends on the age of the snake and the position of the square. As you can see from these skins, the squares are different in size depending on where they are along the body. What is interesting is that the sides of the largest square, on the centre of the rattler's body, are usually 13 scales long. This I believe to be the reason why the Maya regarded the number 13 as sacred.'

We spent some more time discussing his ideas and then, it being late, we made to leave. Before we could do so, he asked me if I would do him one last favour. Taking up copies of *La Serpiente Emplumade: Eje de Cultures* and *Mi Descubrimiento del Culto Crotalico*, his two major works, he pressed them into my hands. 'Please,' he said, 'take these and when you are done with them pass them on to a suitable institution in England. When you get to my age, you feel how death is stalking you and getting ever closer. I would like to know that after I am dead and gone, even if my discovery of the cultural implications of the rattlesnake is ignored in Mexico, my work will be available to serious scholars in England. I would also like you to have these two rattlesnake skins. You may find them of interest.'

With that, we left him and returned by taxi to our hotel.[2]

CITY OF THE PLUMED SERPENTS

The following day, we were up early and off to visit the most famous of all Mayan cities: Chichen Itza. We spent several hours wandering among the buildings of the main plaza, some of which were restored in the 1930s by J. Eric Thompson. After our meeting with Don José, it seemed that everywhere we looked – on walls, on pillars and on the clothes worn by sculpted warriors – there were rattlesnakes. The most prominent of all were the giant upturned serpents, their tails held up in the air, that were paired either side of stairways and entrances to major temples. I couldn't help but notice that the way these tails were positioned vertically, they would cast no shadow at noon on the days of zenith transit. This would have been very noticeable for someone in the know. Clearly Don José was right: the people who built Chichen Itza were very aware of such transits and used them for marking the start of their New Year.

I found confirmation that the Yucatecan Mayans started their year on 16 July in the writings of Bishop Diego de Landa. Although he was writing some centuries after the demise of Chichen Itza, there can be little doubt that the festivals he describes were ancient in origin. Landa tells us that the first month of the year was called Pop and that it began, as José said, on 16 July. This festival was evidently a time of renewal, spiritually as well as physically.

> The first day of Pop, which is the first day of the Indians, was its New Year, a festival much celebrated among them, because it was general and of all; thus the whole people together celebrated the festival for all their idols. To do this with the greater solemnity, on this day they renewed all the service things they used, as plates, vases, benches, mats and old garments, and the mantles round their idols. They swept their houses, and threw the sweepings and all these old utensils outside the city on the rubbish heap, where no one dared touch them, whatever his need.
>
> For this festival the chiefs, the priests and the leading men, and those who wished to show their devoutness, began to fast and stay away from their wives for as long time before as seemed well to them. Thus some began three months before, some two and others as they wished, but none for less than thirteen days. In these thirteen days then, to continence they added the further giving up of salt and pepper in their food; this was considered a grave penitential act among them. In this period they chose the *chacs*, the officials for helping the priest; on the small plaques which the priests had for the purpose, they prepared a great number of pellets of fresh incense for those in abstinence and fasting to burn to their idols. Those who began these fasts, did not dare to break them, for they believed it would bring evil upon them or their houses.

When the New Year came, all the men gathered, alone, in the court of the temple, since none of the women were present at any of the temple ceremonies, except the old women who performed the dances. The women were admitted to the festivals held in other places. Here all clean and gay with their red-coloured ointments, but cleansed of the black soot they put on while fasting, they came. When all were congregated, with the many presents of food and drink they had brought, and much wine they had made, the priest purified the temple, seated in pontifical garments in the middle of the court, at his side a brazier and the tablets of incense. The *chacs* seated themselves in the four corners, and stretched from one to the other a new rope, inside of which all who had fasted had to enter, in order to drive out the evil spirit . . . When the evil one had been driven out, all began their devout prayers, and the *chacs* made new fire and lit the brazier; because in the festivals celebrated by the whole community new fire was made wherewith to light the brazier. The priest began to throw in incense, and all came in their order, commencing with the chiefs, to receive incense from the hands of the priest, which he gave them with as much gravity as if he were giving them relics; then they threw it a little at a time into the brazier, waiting until it ceased to burn.

After this burning of the incense, all ate the gifts and presents, and the wine went about until they became very drunk. Such was the festival of the New Year, a ceremony very acceptable to their idols.[3]

Bishop Landa's account makes no mention of the priests watching shadows cast by the Sun. However, I checked the choice of date for the New Year with the Skyglobe computer program. Sure enough, 16 July is the day when the Sun crosses the zenith over Chichen Itza, Mani, Merida and other places in the northern Yucatán. It was clearly a solar festival in celebration of the Sun's zenith transit.

Something of the excitement attending such a festival was evident on the day we arrived, too. It was 21 March, the spring equinox, and a large crowd of 30,000 or more was gathered in the park surrounding the buildings. Unfortunately, the principal monuments, the Castillo, or Pyramid of Kukulcan, the Temple of the Warriors and the Caracol, were all roped off and we were unable to climb them. However, that didn't seem to dampen the enthusiasm of the thousands of gathered visitors. Many were Native Americans, dressed in brightly coloured costumes, with elaborate headdresses made from feathers. Near to the pyramid there was a makeshift stage on which, in turn, groups of costumed performers danced to the music of pipes and drums. Though the dances were of modern invention and no more traditional than the aniline dyed cloth and sewn on sequins that made their costumes so striking, they nevertheless gave a glimpse of the past. I could well imagine similar dancers performing a millennium earlier on the raised up stage today called the Platform of Venus.

As the afternoon wore on, so the crowd moved closer to the pyramid. The object of everyone's attention was the balustrade of the staircase on the northern side of the pyramid. At first, the sky was quite overcast with clouds, giving welcome protection from the Sun's rays, which at this latitude, 20° 40′ North, are strong even at the beginning of spring. Unfortunately, although the shrouding of the Sun behind clouds made the temperature more bearable, it also obscured what we had come to see. Eventually, almost at the last minute, there was a break in the clouds and a shadow appeared on the north-western side of the northern staircase. Because of the shape of the pyramid, it gave the appearance of an undulating serpent that linked the dragon-like head at the bottom of the stairs with the temple on top of the pyramid. When the shadow appeared, a great cheer went up from the crowd. People blew whistles and I just had time to take a photograph of the phenomenon before the Sun went back behind the clouds.

It seems that the inhabitants of Chichen Itza – partly Mayan, partly Toltec – deliberately oriented the axis of the pyramid some 19° away from true north for just this effect. However, having now seen it for myself I felt somewhat cheated. My friends and I had travelled many thousands of miles to witness this event, which to be honest was a bit of a disappointment. I had to remind myself that, though the appearance of the shadow-serpent on the side of the staircase might not seem so special to me, to a Mayan farmer 1,000 years ago it must have seemed extraordinary. Also, if it was true that the pyramid was deliberately oriented in such a way that it would perform this miracle of shadows, then that in itself was exciting. It meant the pyramid was a silent witness to the knowledge of the Mayan priesthood and symbolised their confidence that the Sun could be relied upon to cast such a shadow year after year, katun after katun. Clearly, this play of shadows must have been of considerable comfort to people whose religion taught that this was anything but guaranteed. That their priest-astronomers were able to work out that a pyramid of this shape, oriented 19° away from true north would do this, is an indication, if any is needed, of their skill as astronomers. The question remained, though, what might be the reason for orienting the pyramid the way it was?

THE SERPENT AND THE ZENITH

Surprising as it may seem, I believe that the answer to this question has much to do with zenith calendrics. That the pyramid was intended to be a solar monument is not in question. It has four staircases leading to its top, one running up the centre of each face. Each staircase has 91 steps, making a total of 364. If we add on the extra step leading into the sanctuary on top, this makes 365: the number of days in the Haab cycle, or year. This number is reminiscent of the number of stairs on the Pyramid of the Sun at Teotihuacan, which I climbed only a few days earlier: again 365. This must have been deliberate, but I believe the pyramid had a purpose other than art: it was a means of calibrating the true length of the year and thereby keeping their calendar in step with reality.

The true length of the solar year is nearly 365.25 days long. To take account of this extra quarter day and to keep our calendar more or less in sequence with the Sun's movements, we add on an extra day every fourth or 'leap' year. The Maya were up against the same problem with measuring time. In order to maintain the integrity of their months so that their new year, 1 Pop, would begin with the zenith transit of the Sun, they needed periodically to add on extra days. They don't seem to have done this every four years, as we do. However, there is some evidence that at the end of a 52-year cycle (what the Aztecs called a 'sheaf' of years) they would use the opportunity to reset their calendar by adding on 13 days. If this is the case, then maybe one purpose of the pyramid was to reassure the people that this was a correct thing to do. Seeing the equinoctial shadow appear on the correct day – 116 days before the start of a new year – could be taken as a sign from the sun god that the priests were doing the right thing by adding 13 extra leap days to the calendar. There is, however, something else that has to be considered. The pyramid of Kukulcan is not just a sundial: it is also deeply symbolic.

This was something that Don José explained to me when I first met him in 1994. The design of the Pyramid of Kukulcan was intended to symbolise a coiled up rattlesnake. Just as a rattler twists itself in coils that are of ever-decreasing circles, so the pyramid rises in nine tiers. The head, and therefore the mouth, of a coiled rattler is positioned at the centre and top of its coiled body. So too is the sanctuary placed on top and at the centre of the pyramid, its open 'mouth' facing roughly north. The plan view of the pyramid mirrors the Canamayte pattern on the serpent's back. Viewed from the ground, the walls of the pyramid are faced with stone blocks that replicate the scales of a snake's skin. Thus it is that in an extraordinary and abstract way, the pyramid of Kukulcan, or Quetzal-serpent, symbolises a coiled up rattlesnake.

I was quite surprised to hear this and not a little sceptical when he told me it. However, after reading what today's Mayanists had to say on the subject of pyramids I could see the truth in what he said. According to Linda Schele and David Freidel, writing in their now classic text *A Forest of Kings*, Mayan pyramids were intended to represent mountains and often the temple on their top symbolised a gateway leading to the 'other' world. A group of temples represented a mountain chain.

A Forest of Kings was published in 1990. By the time Schele died (in April 1998), her ideas on pyramids as monster mountains had progressed further. In her last book, *The*

Code of Kings, she describes 'Snake Mountain' as a mythic place of origin for mankind. She says that, according to Aztec myths, at the foot of the original Snake Mountain – *Coatepec* in the Aztec language, *Chan-witz* in Classic Mayan – was a ball-court. Near this was a lake or well, and here the first city was built. It was called Tollan in the Aztec language and its inhabitants were the original *Toltecs*. In retrospect, they were highly regarded as having been the inventors of civilisation at the start of the present age. More than that, Tollan itself was the archetypal city on which all later cities, e.g. the legendary Teotihuacan and the Aztec's own capital of Tenochtitlan, were modelled. Such ideas were not confined to the Aztecs, for it seems that the Mayans too modelled their cities after this archetypal city, which in Classic Mayan was called *Puh*. Consequently, they too graced their cities with 'snake mountain' pyramids, ball-courts and wells.

It is easy to see that Chichen Itza has all the hallmarks of a 'Tollan' city. The pyramid of Kukulcan is its 'Snake Mountain', close by it is a ball-court – the largest in the whole of Central America – and not far from this is a *cenote*, or well, upon whose waters the life of the city depended. It could be argued that these features result from the 'Toltec' influence supposedly deriving from the invasion of the Yucatán by people called the 'Itzas' in c. AD 850. Yet it is clear from other sites, such as Uxmal, Tikal and Palenque, that the arrangement of serpent mountain(s), ball-court(s) and sacred well(s) is common to Mayan as well as Toltec/Aztec cities. For the Maya, like the Aztecs and Toltecs, pyramids were symbolic of mountains. Why it was considered necessary to build cities in imitation of Coatepec and Tollan is not clear. The clue, however, like so much else, lies in the *Popol Vuh*. What happened at 'Serpent Mountain' is not clear but the archaeological remains of pyramids tell us a certain amount about its significance. These pyramids usually contain some sort of chamber, either within them or in the form of a sanctuary on top. In the case of the Kukulcan pyramid at Chichen Itza, there are two sanctuaries: one is easily accessible on top of the pyramid and the other lies within. Access to the sanctuary on top is guarded by two enormous stone rattlesnakes, their dragon-like mouths projecting from the foot of the north side. The question is, does anything lie behind the myth of Snake Mountain as a place of creation?

This is not an easy question to answer, but I believe a possible clue lies again in the *Popol Vuh*. In the later section of this work, after the story of Seven Macaw and the Hero Twins, we are told of what happened at the beginning of the age. Darkness enveloped the Earth while the people, many tribes, gathered at *Tollan Zuhuya*. This place, with 'seven caves and seven canyons', was in the mountains and therefore safe from flooding. Here they were told, presumably through their shamans, that they would receive their gods. Unlike the great creator gods who represented the intangible forces of nature, these tribal gods were more like familiar spirits. They were closely linked to idols made of wood or stone that the tribes subsequently carried around in knapsacks.

In the *Popol Vuh*, the three most significant of these idols are named *Auilix*, *Hacaulitz* and *Tohil*. The last of these was the most important of the three, at least as far as the Quiché were concerned. He saved them when they were shivering in the cold and darkness by giving them the gift of fire. This, however, came at a cost. The tribes who received this boon had to take an oath that in due course their descendants would 'suckle' the god: a euphemism for giving blood sacrifices.

Now Tohil, the titular deity of the Maya, was also known by other names: *Tahil*, *Bolon Tsacab*, Tezcatlipoca and, most notably, *Kawil*. Under this last name, he is met with at Palenque, where he was also referred to by archaeologists as either God K or GII. Often, Kawil is shown as a dancing figure with a celt, or stone axe, penetrating his forehead. Why this should be, we don't know, but axes were also symbolic of thunderbolts. As the dead Pacal is similarly shown on his tomb lid with the axe of Kawil buried in his forehead, it could be that being thus smitten is indicative of death itself.

Kawil was also a serpent god, for, although he has a human body, one of his legs terminates in a serpent's head. In the *Popol Vuh*, when the fires of the people have gone out, Tohil (the Quiché name for Kawil) spins around and around in his sandal like a primitive stick-in-block firelighter. This is how he generates the fire he subsequently shares with his people. His link with fire is also indicated by the fact that he is frequently shown holding a lit cigar.

Throughout the Mayan civilisation, and maybe from long before, Kawil is the deity most especially associated with kingship. The lords of Palenque (and other Classic cities) claimed their authority from Kawil, who, provided they reciprocated with appropriate sacrifices, acted as the patron of royalty. Worship of Kawil took the form of renewing their contract with him. For this, it was necessary that they open a symbolic 'portal' linking the earthly sphere with the heavenly. The way the Maya lords did this was to cut themselves and allow their own blood to flow into a shallow bowl, where it was absorbed onto paper. Men would normally draw blood by piercing their penises with sharp implements. Women would bleed themselves from the tongue by threading through it a string bearing sharp barbs. In both cases, the extraction of the blood must have been extremely painful and not a little frightening. Once the paper had dried, it was set on fire along with other materials thought to contain *k'ulel*, or lifeforce. The smoke from the fire would rise up and, if the ritual was carried out correctly, form into a 'vision serpent'. Out of the mouth of this serpent, depicted on lintels such as the one now in the British Museum, ancestral spirits were able to emerge and give guidance to the king. The king evidently needed to be a trance-medium or medicine-man to hear them.

Besides his primary symbol of the manikin sceptre, Kawil was also associated with the double-headed serpent bar. According to Schele, this represented the zodiac, the pathway which the Sun and planets travel along. There is some controversy as to what exactly the dragon heads at either end of the serpent bar represented. According to some Mayanists, they symbolise the places in the sky where the Sun crosses the celestial equator – which it does on the equinoxes. At first sight, this seems an attractive idea. However, other than the appearance of the sliding shadow-serpent on the side of the staircase at Chichen Itza, there is little indication that the Maya paid much attention to the equinoxes. I believe what they were more concerned with were zenith transits of the

Sun, the second of these[4] being used as the start of a new year. The importance of this second zenith transit, as Don José Diaz Bolio said, is due to the fact that rattlesnakes shed their skin around this time. They do this by emerging out of the mouth of the old skin, leaving this behind them like a discarded shell. A snake's head is therefore a portal of rebirth. Thus, gods and ancestors alike are frequently shown emerging from such a head.

Figure 44: Eagle warrior from Chichen Itza embraced by vision serpent and clouds of sacred smoke

Taking all this information together, it became clear to me that the July zenith transit of the Sun was looked upon by the ancient Maya as its rebirth. This was reinforced when, after I returned home, I began to explore the significance of dates recorded at Palenque.

PALENQUE, THE PLACE OF SERPENTS

According to Father Ordoñez, Palenque was known to the ancients as *Nachan*, or Place of Serpents. Whether this was so in later days, we don't know. However, the now-translated hieroglyphs tell us that in Classic times Palenque was called *Bak*, or bone, and that it was the capital of a principality called *Lakam-Ha*, or Great Water. Nevertheless, the city abounds in serpentine symbolism, not least the serpent bars on the lid of Pacal's sarcophagus. Could it be, I wondered, that Ordoñez's Place of Serpents was more in the nature of a nickname, rather as New York is known as the Big Apple, Chicago as the Windy City and London as the Big Smoke? If that were so, then why might it have acquired this name?

I believe that the answer to this question is quite simple: the word Nachan has been mistranslated. In many Mayan languages, the word chan (or *kan*) has two meanings. It can mean 'snake' but it can also mean 'sky'. The Mayan scribes were fully aware of this, for in their hieroglyphs they frequently used the homophonic glyph for serpent when they meant to say sky. This being the case, it could be that the proper translation of Nachan is not 'Place of Snakes' but rather 'Sky City'. This would fit rather well with Palenque. Anyone who has been there will appreciate how the view from the top of the Pyramid of Inscriptions gives an unrivalled view over the plains to the north of the city and also to a wide expanse of sky in this direction. What is also true is that the so-called 'Palace' housed what, for want of a better term, we can call an observatory. Around the cornice in certain rooms can still be seen moulded hieroglyphic symbols of the Moon and certain planets. Meanwhile, the tower, perhaps the most striking feature of this building, could have been used for observing zenith transits.

We may never know whether 'Sky City' is how the ancient

Maya referred to Palenque but we can be pretty sure that it was home to a school of astrologers. We know this because among the many inscriptions is the most authoritative collection of astrological inscriptions in the whole of Central America. These are relatively well preserved and accompanied with large quantities of texts and dates.

Most of these are contained in the remarkable collection of buildings at Palenque today known as the 'Group of the Cross'. These temples, a 'symbolic mountain range', were constructed by King Chan-Bahlum II, the son of Pacal the Great. They were built around AD 684 and contain important astrological data. Besides giving us details of his own lineage, Chan-Bahlum gives us the date of the start of the present age (13 August 3114 BC) and links this with the birth of three important gods of Palenque, who Mayanists have long nicknamed GI, GII and GIII. God GII, as we have seen, is now known to have been the same personage as Kawil or Tohil, the deity who, according to the *Popol Vuh*, gave fire to mankind at the beginning of the present age. Gods GI and GIII correspond to the Hero Twins: GI being the equivalent of Hunahpu, the shooter of Seven Macaw, and GIII his jaguar-spotted brother Xbalanque. According to Schele and Freidel, GI was also associated with the planet Venus, while his brother was the 'cruller-eyed' jaguar god, who is also known as Ahau-Kin, 'Lord Sun'.

I wasn't sure how this fitted in with the larger picture of sacred astronomy at Palenque, but as Schele and Freidel published a number of dates associated with these gods, I decided to look into the matter. The dates of the 'births' of these gods are given (as are other important dates) in the buildings of the Cross Group of Temples. I checked these dates with my computer and found, to my surprise, that almost invariably they corresponded with important astral events involving either the Moon or the planets Mercury, Venus, Mars, Jupiter and Saturn. It became clear from this that, in addition to any link with Venus or the Sun, GI was an avatar or personification of Saturn, GII of Jupiter and GIII of Mars.

STAR NOTES FROM PALENQUE
(Based on translations given by Schele and Freidel in *A Forest of Kings*.)

Temple of the Cross inscriptions
1) *'On 7 December 3121 BC, Lady Beastie the First Mother was born.'*
   Commentary: Jupiter transits the zenith just before sunrise. Moon and Venus close together and visible at dawn. (The actual conjunction took place the afternoon before and would not have been visible.)

2) *'On 16 June 3122 BC, GI', The First Father was born.'*
   Commentary: There was a very close conjunction of Saturn with Mercury on this day and both transited close to the zenith (within 2°) about an hour before noon. This, however, would not have been a visible event. The Sun also transited within 2° of the zenith.

3) *'On 13 August 3114 BC, the 13th baktun ended and the new creation began.'*
   Commentary: Mercury transited exactly at the zenith. Saturn also transited (about an hour and a half later) about 1° from the zenith.

4) *'On 5 February 3112 BC, GI' entered the sky and he dedicated the house named "wacah chan xaman waxac na GI'" [the "World-tree House of the North"].'*
   Commentary: Saturn transited about 10° from the zenith as the Milky Way rose to form an arch in the sky: the World-tree House of the North.

5) *'On 21 October 2360 BC, GI, the child of Lady Beastie, was born.'*
   Commentary: The Moon, in its last quarter, culminates exactly at the zenith shortly after sunrise. It would probably still have been visible as it did so. About three hours later, Venus transits close to the zenith (about 1° away).

6) *'On 13 August 2305 BC, at age 815, Lady Beastie became the first being in this creation to be crowned as king.'*

Commentary: Mercury transited the zenith point, but this would not have been seen because it was close to the Sun.

7) *'On 11 March 993 BC, U-Kix-Chan was born.'*
Commentary: Mars transited 2° from the zenith.

8) *'On 28 March 967 BC, at age 36, U-Kix-Chan, Divine Lord of Palenque, was crowned king of Palenque.'*
Commentary: The Moon culminates exactly on the zenith, though this would occur in the afternoon and therefore not be visible.

9) *'On 31 March AD 397, Kuk was born.'*
Commentary: Venus culminates about 1° from the zenith. (It would be exactly on the zenith three days later.)

'It was 22 years, 5 months, 14 days after he had been born and then he crowned himself on 11 March AD 431. He was Divine????? Lord.'
Commentary: Venus transits exactly at the zenith on this day.

10) *'On 9 August AD 422, Casper was born.'*
Commentary: On this day, the Sun culminates close to the zenith.

11) *'13 years, 3 months, 9 days after Casper had been born and then it was 10 August AD 435, 123 days after Casper crowned himself.'*
Commentary: There is a tight conjunction of the Sun and Moon on 10 August AD 435 that may have meant a partial eclipse visible in Palenque. This was some four days before the Sun made its zenith transit. One hundred and twenty-three days before this is 9 April 435. On this day, Jupiter culminates about a degree from the zenith. There is also a conjunction of Venus and Mars. None of this can been seen, though, as it happens in daylight.

12) *'And then 11 December AD 435 came to pass, on that day 3,600 years (9 baktuns) ended.'*
Commentary: Jupiter is placed at the star-gate over the hand of Orion. It culminates about 5° from the zenith point.

13) '42 years, 4 months, 17 days after he had been born and then Chaacal-Ah-Nab crowned himself on 4 May AD 565.'

Commentary: On this day, the Sun comes close to transiting the zenith. It actually does so two days later, on 6 May AD 565. Also, on 4 May, Saturn was transiting the zenith point at sunset.

Temple of the Foliated Cross inscriptions

1) 'On 8 November 2360 BC, when the eighth Lord of the Night ruled, it was ten days after the Moon was born, five Moons had ended, X was its name and it had thirty days . . . It was the third birth and GII was born.'

Commentary: The Moon is exactly in conjunction with the Sun (and therefore invisible) on 28 October. The new moon is 'born' the following day, on 29 October. Counting on ten days does bring us to 8 November, so the record is accurate! The following day, the Moon would have conjuncted Saturn.

2) '34 years, 14 months after GII, the Matawil, had been born and then 2 baktuns (800 years) ended on 16 February 2325 BC. On that day, Lady Beastie, Divine Lord of Matawil, manifested a divinity through bloodletting.'

Commentary: The Moon passed within 2° of the zenith just as the Sun set.

3) 'On 8 November 2360 BC, GII, the Matawil, touched the Earth . . .'

Commentary: The Moon is below the Pleiades, fairly close to Saturn.

'3,050 years, 63 days later on 10 January AD 692 . . .'

Commentary: On the night before this, around 9.30 p.m. on 9 January AD 692, the Moon transited exactly over the zenith point.

'. . . on 23 July AD 690, GII and GIII were in conjunction.'

Commentary: Jupiter and Mars were in conjunction on this day. Saturn was close by.

4) '46 years, 6 months, 4 days after he had been born and then he crowned himself, Lord Chan-Bahlum, Divine Palenque Lord on 10 January AD 692.'

Commentary: The day before this, the Moon transited the zenith.

From all of the above, it is clear that the Maya computed the movements of the planets very accurately. The data shows that, to the Maya, Lady Beastie, the First Mother, was an avatar of the Moon. Lady Beastie is called by this name because her hieroglyph, while identifiable, has not yet been translated. She is associated with the Moon but also clearly equates with the girl in the *Popol Vuh* story who is made pregnant by the head of Hun Hunahpu, or First Father.

Finding the planetary identity of the other gods, GI', GI, GII and GIII, was less obvious. What was clear was that significant events involving these gods were timed to coincide with portents involving the planets, especially Jupiter, Mars and Saturn. There was no surprise in this as these planets are among the brightest objects in the sky and we know their cycles were closely followed by the Maya. What was more curious was the evident importance attached to zenith transits, something not much mentioned in the books I had read. This, however, did make sense for a people used to living in jungle conditions. For them, the passage of a planet across the zenith point, directly overhead, would be much more easily observed than either its rising or setting. It would seem from this that Chan-Bahlum's rituals involved invoking the appropriate god when his planet was directly overhead. The zenith point, therefore, was the birthplace of not just the Sun but the planets too.

To test this theory further and make sure this was not just a personal fetish of Chan-Bahlum's, I looked at one other date that Schele and Freidel give. This was the accession of King 6-Cimi-Pacal on long-count date 9.18.9.4.4. This translates to 15 November AD 799 on the Gregorian calendar and is the last dated inscription at Palenque. Feeding this information into my computer, I was able to see that on this day, at around 3.27 a.m., the planet Saturn culminated over Palenque exactly at the zenith. The fit was far too exact for it to be a coincidence. It implied that the Maya of Palenque were still tracking the planets and arranging important rituals to take place on zenith transits right up until the abandonment of their city shortly afterwards.

The role of GII, or Kawil, the god of royalty, I found particularly significant. It was clear that he was symbolically linked with the planet Jupiter. This I found very interesting, as Jupiter is also the royal planet par excellence for our own civilisation. Not only that, but Jupiter's Greek equivalent, Zeus, has mythological connections with the story of Prometheus bringing fire to mankind. Also, Zeus/Jupiter, like Kawil, was associated with lightning and thunderbolts: both activities that can lead to the starting of fires.

As we have seen in the previous chapter, Zeus and his antagonist Prometheus are also very much connected with the myth of Atlantis. The great flood of Deucalion appears to be a myth concerning the destruction of Atlantis. It was after this that Prometheus stole a glowing ember from Zeus's hearth and gave it to mankind. If the destruction of Atlantis was a real event, as Edgar Cayce claimed, then the Maya myth concerning Kawil's fire is an equivalent story. Pursuing the connection between these Greek and Mayan myths further would bring me to the final stage of my quest.

CHAPTER 11

At the End of the Age

Returning to Britain, my journey at an end, I had much food for thought. Everywhere I visited underlined the precariousness of civilisation. The ruined cities of Chichen Itza, Palenque, Uxmal and Comalcalco were stark reminders that, however advanced they might seem, sooner or later even advanced civilisations come to an end. With the dire consequences of global warming all too apparent and Edgar Cayce's predictions of sudden and catastrophic Earth-changes in the near future, there is much for us to be concerned about with regard to our own civilisation.

Accordingly, I was motivated to take a fresh look at the Mayan prophecy for the end of the present age. I wanted to see how well this accorded both with observable astronomy and with the changes we see going on around us. Above all, I was anxious to know of any signs in the sky that might give us some warning of what might happen on the fateful day of 22 December 2012.

Repeated reference has been made throughout this book to the Mayan start-date of 13 August 3114 BC. As we have seen, zenith-point astronomy indicates a correlation between the *Popol Vuh* story of the shooting of Seven Macaw and the zenith transit of the constellation of *Sagitta* and the wounded wing of Aquila, the eagle constellation over

Teotihuacan. However, this transit was occurring daily for a number of years, while the calendar is much more specific than that: it points to 13 August 3114 BC as the exact day on which the counting of days started.

Using the same computer program as before and setting this to the date in question, I could not see anything in the sky that was of such importance that it would have caused the Maya (or anyone else, for that matter) to have regarded this day as special. Unless something physical had happened on that day, such as the eruption of an important volcano, the founding of an empire or the appearance of a prophet, there seemed little reason for choosing 13 August rather than 13 June, July, September or October. The choice of a date in August seemed arbitrary.

The Maya, however, did not only define time periods in their calendar by counting the number of days that had gone by since 13 August 3114 BC. They also counted backwards, from long-count date 13.0.0.0.0 – 4 Ahau 8 Cumku – when the 13th baktun would be complete and their day-count would go back to zero. As this end-date, which we compute to 22 December 2012, corresponds to the winter solstice, this is obviously important. However, astronomically speaking, it is of much more significance than this. Not only is the night of 21–22 December the longest in the year, but because of the precession of the equinoxes it at present corresponds with the day the Sun stands exactly at one of the 'star-gate' crossing-points of the ecliptic with the median plane of the Milky Way.[1] I have named this position the southern star-gate – its counterpart, the northern star-gate, being placed exactly over the upstretched hand of Orion. As I have said, the southern star-gate is of principal importance because it is also aligned with the centre of our galaxy. What this means in practice is that on 22 December, any person observing the Sun will also be looking directly towards the core of the Milky Way: the place where astronomers say there is a black hole with a mass some three million times that of our Sun.

Now it could, of course, be a simple case of coincidence that the Mayan calendar reaches the end of its 13th baktun on precisely this day. However, if this is a coincidence, then it is certainly extraordinary. What it seems to indicate is that

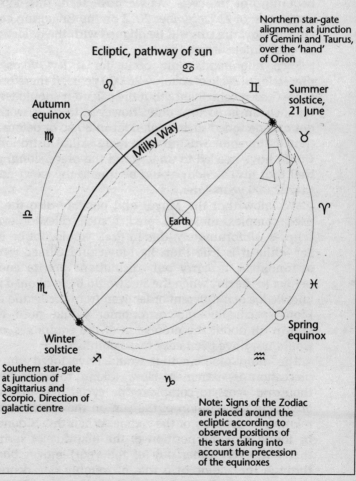

Because of the precession of the equinoxes, the sun presently aligns with the 'star-gate' positions (where the median plane of the galaxy intersects with the ecliptic) at the solstices
The Mayan long-count calendar reaches its conclusion on 22 December 2012 with the sun aligned with the galactic centre at the winter solstice

Northern star-gate alignment at junction of Gemini and Taurus, over the 'hand' of Orion

Ecliptic, pathway of sun

Summer solstice, 21 June

Autumn equinox

Milky Way

Earth

Spring equinox

Winter solstice

Southern star-gate at junction of Sagittarius and Scorpio. Direction of galactic centre

Note: Signs of the Zodiac are placed around the ecliptic according to observed positions of the stars taking into account the precession of the equinoxes

Figure 45: Alignment with the southern star-gate

the Mayan long count was devised with the future, not the past, in mind. Whoever started writing long-count dates was not so much concerned with the past as with the future. Although the calendar begins with a notional date of 13 August 3114 BC, we have no recorded dates of anything like this age. The earliest date we have is Olmec and is from 30 BC. The real purpose of the calendar seems to be to place dates in relation to a future date at the end rather than the beginning of the cycle. As we have seen, this date is the equivalent of 22 December 2012 on the Gregorian calendar. On this day, the sun will be aligned with the galactic centre on the winter solstice.

Such alignments only occur for a few years in the precessional cycle of roughly 26,000 years. It therefore seems more than a coincidence that the calendar should terminate on such an important date. However, to have worked this out and designed such a calendar in 30 BC or before implies that its inventor was an extremely skilful astronomer. He would have needed to understand the precessional cycle so well that his calendar would be true to the exact day more than 2,000 years later.

We know that the Maya, and possibly also the Olmec, used complex tables to predict such celestial events as eclipses. Unfortunately, this in itself would not be enough. For while it is true that the Maya and Olmec were good astronomers, to carry out calculations precise enough to predict to the day when the Sun would be so aligned requires knowledge of a different order even from predicting eclipses. More specifically, our astronomer would need to have known all about the precession of the equinoxes, and this would not have been easy to accomplish.

The main problem that would have faced him is that precession is extremely slow, taking roughly twenty-six thousand years to complete one cycle. It is measured by observing the position of the Sun on the spring equinox in relation to the stars of the zodiac. When this is done, it can be found that the position of the first day of spring (and therefore every other day of the year) moves backwards through the zodiac at a rate of roughly one degree every seventy-two years. To discover this requires not just precise observation by an alert individual but careful record-keeping extending over generations. As we have no evidence that the Olmecs had more than a very rudimentary written language, it is difficult to see how these records were kept. As the Greeks discovered and wrote about precession in *c.* 130 BC, it is just possible that whoever wrote the first long-count date learnt about precession from this source. However, this too is not without its problems. Even if an Olmec priest living in 30 BC was able to read Greek and learned about precession in this way, he would still have great difficulty in computing the correct number of days for the count to end on a day when the Sun was aligned with the star-gate and galactic centre on the winter solstice. Yet, difficult as this may be, we have the evidence before us from countless inscriptions that this level of accuracy was achieved. The question is how?

One of the most curious things about Central American civilisation is that it has so much in common not only with the earlier Egyptian but also the Chinese. This has been remarked upon by many scholars, past and present. It is noticeable, for example, that both the Chinese and the Maya had a great liking for jade. In both cultures, this precious stone was used for grave-goods. While this in itself might not seem strange, it is when one realises that it was the custom in both places to put jade beads in the mouth of the deceased. Not only that, but the custom of supplying jade masks for Mayan kings is strangely reminiscent of the way Chinese emperors were buried in jade suits.

These are not the only similarities. It has been shown that several of the day-name signs used by the Maya are exact analogues for named days of the Moon as used in China and throughout south-east Asia. It has also been discovered that the method used by the Maya to calculate solar and lunar eclipses was exactly the same as that employed by the Chinese of the Han Dynasty (*c.* 200 BC–AD 200). Yet despite the fact that much of the sculpture found at Teotihuacan and elsewhere has a curiously Chinese look to it, not a single Asian-sourced artefact has yet come to light in the archaeology. There is also the curious anomaly (if we are talking about cross-cultural contact) that Mayan society was still in the Stone Age and had no wheels at a time when the Chinese, for whom wagons of all sorts were ubiquitous, were already highly proficient in the use of not just bronze but iron. This lack of firm evidence for contact has hardened the scepticism of archaeologists against the idea of any sort of trans-oceanic cultural contacts. But is this justified?

In an earlier chapter, we discussed the possibility, which in my opinion is very likely, that Mexico was visited by Europeans many centuries before Columbus and even before the epic voyages of the Vikings. Scattered remains of Roman coins, amphorae and even a small sculpted head suggest that there was maritime contact in at least the fourth century AD. The presence of Carthaginian scripts and possible connections between the Olmecs of Tabasco and the Mandeans of West Africa push cross-Atlantic contacts back even further to maybe 300 or 400 BC. Even this, though, does not seem to be the date of the first transatlantic contacts. The evidence of the Davenport Calendar Stone would seem to indicate that North America was visited by people from the late-bronze to early-iron age, c. 1000–500 BC. However, this is a very conservative estimate. It is certainly not beyond the bounds of possibility that people were voyaging to America from as early as 3100 BC. Could they have told the people of Mesoamerica about the cycle of precession?

The taboo against discussing transatlantic contacts, which represents political correctness at its most insidious, once extended to North America as well. However, in the United States and Canada, while there is still denial of the possibility of any Roman or Carthaginian links, there is now acceptance that Viking explorers probably did make the voyage to America – as indeed their sagas claim. Furthermore, very recent work at many sites in both North and South America is undermining the assumption that all native Americans prior to the arrival of Columbus are descended from Alaskan immigrants who crossed the Bering Straits. Archaeological evidence now indicates that South America was settled prior to North America, which implies that America's earliest settlers came by sea.

However, this is not the end of the matter. Throughout the Maya lands it was the custom to strap the head of a newborn baby into a wooden press. This was kept done up tightly for some months until the cranium hardened. The purpose of this seeming torture, which must have caused great distress to babies and parents alike, was to flatten the forehead, thereby giving a streamlined look known as a *polcan*, or snake-head. Why, one must ask, did they do this? Surely babies would not be put through this torment without good reason?

Why too did the lords of Palenque, Yaxchilan, Bonampak and other cities wear false noses? Was this too a matter of fashion or did large Caucasian or Semitic noses indicate a special status? That this may have been the case is indicated by another curious detail: the Mayans practised circumcision. This seldom-mentioned fact is indicated very clearly on some extant statues. I have seen statues of circumcised men at Dzibilchaltun and Kabah, both cities in the Yucatán region. These men were prisoners who had been stripped bare as part of their humiliation, but it seems likely that circumcision was widely practised, at least among the nobility.

A further curiosity is the extraordinary stature of some nobles (notably Pacal the Great) compared with the ordinary populace. The skeleton of Pacal, which lies buried in his sarcophagus at Palenque, was examined in detail after its discovery in 1952. He was found to have been a very tall man of about 6 ft, which means that like me (6 ft 2 in.) he would have stood head and shoulders above the average Mayan. Since then, at least one other skeleton of a similarly sized giant has been found at Palenque, indicating that Pacal himself was not a freak but a member of a family or family group with the genetic trait of tallness.

When we put all these pieces of evidence together, a picture begins to emerge that is quite at variance with what is currently taught in the schools of academic Mayanology. We can surmise that the reason that having a polcan, or flattened head, was considered essential for a nobleman is that at one time this was a recognisable genetic trait of the ruling class. So too might have been a tall stature and having a large nose. Over a period of time, through successive marriages with people outside the immediate family group, some of these traits may have been lost. While little could be done about the height of an individual, it was possible to flatten the head of a baby to give the appearance of 'natural' nobility. Also, for ceremonial purposes at least, a false nose or even false beard could be worn.

As has been said, having a large nose is typical of some Europeans. However, prior to the last century, few Europeans would consider having either themselves or their sons circumcised. This is essentially a Middle Eastern and African custom. Historically, it is closely associated with the religions of Judaism and Islam, though it is also now widely practised in America, for non-religious reasons. Yet it was not the Jews who invented circumcision. Long before the time of Abraham, circumcision was being practised in Egypt. We know this from documentary evidence but it is confirmed by a statue in the Old Kingdom rooms of the Cairo Museum showing a naked male who, like the unfortunate prisoners at Dzibilchaltun and Kobah, has clearly had the operation.

The Egyptian or African connection is very interesting for other reasons. A flattened forehead and prominent chin is a trait of certain African tribes to this day. As is well known and recorded in all the history books, the XVIIIth Dynasty, which brought about the expulsion of the Hyksos, came from southern Egypt. One of its most famous pharaohs was the heretic king Akhenaten, uncle of Tutankhamun. Busts and reliefs of Akhenaten's wife Nefertiti, who was undoubtedly a great beauty, show that she too had a sloping-back forehead: a natural version of the Mayan polcan. This trait, which must have been familial, was even more pronounced in their daughters, statues of whom can also be seen in the Egyptian Museum.

Now I am not saying here that I think that descendants of Akhenaten's family migrated from Egypt to Mexico, but there could be a more distant and generalised connection. Clearly, at least some people of noble status in Egypt had the genetic trait of elongated flat heads. During the period of the XVIIIth Dynasty, they built an empire that embraced parts of Libya as well as the Levant. Phoenicia, from whence came the later Carthaginians, was thoroughly Egyptianised. However, the Phoenicians, being Canaanites, were of Semitic stock. It is therefore not too fanciful to suggest that, like the Arabs of today, they would probably have had relatively large noses – at least as compared with the Maya.

Carthage itself was founded in *c.* 850 BC but reached the height of its power around 220 BC, shortly before its destruction by the Romans in 202 BC. Like their forebears the Phoenicians, the Carthaginians were a seafaring nation whose ships plied not just the Mediterranean but also the Atlantic. They maintained a monopoly on sea movements to the west of the islands of Sardinia and Corsica so that their customers in the eastern Mediterranean would not have access to their suppliers beyond the Pillars of Hercules. However, it is known that they themselves sailed up the west coast of the Iberian Peninsula and to Britain. They also sailed down the west coast of Africa and may well have rounded the Cape of Good Hope. From the British Isles, they brought back tin and from Africa ivory, slaves and other precious commodities.

The Carthaginians also colonised the Canaries and Azores, which implies more than it seems on the surface. Although these are rather barren islands and of little use in themselves, they would have been invaluable stopping-off points for anyone seeking to cross the Atlantic. From the Canaries, the westerly trade winds could have been picked up. With their aid and assuming there were no major storms or other mishaps, it would have been easy to sail across the Atlantic to South America. From there, it would be a matter of sailing along the coast to get to Mexico. Making the return journey would involve picking up the easterly trade winds. These would take them from the east coast of America to Bermuda, the Azores and back home to the Mediterranean.

That such journeys were clearly possible should not be in doubt. The remains of Bronze Age vessels of a type and strength able to sail the rough seas around the British Isles have been found buried in the mud at Dover and in the Humber Estuary. There is no reason to think that Carthaginian ships of 1,000 years later were in any way inferior. They were probably not much different to those of the Gauls as described by Caesar in 55 BC. These were sailing ships made of oak and, with their high bows, were clearly designed for sailing out in the ocean and not just around the coasts of Gaul.

Unfortunately, the Carthaginians kept their voyages to America a closely guarded secret, most especially from the Romans. The reason they were coy about telling their

Mediterranean customers where they had been was to protect their trade in gold and silver. Even so, we may have learnt more about such journeys had not the Romans, in 202 BC, burnt Carthage to the ground, along with all of its libraries. Though the city was rebuilt, in Roman style, details of the history of the earlier Carthaginian civilisation were almost entirely lost. Carthage, till its destruction Rome's greatest rival, was doomed to become a footnote in history.

The period *c.* 3100 BC is very interesting from a cultural point of view. This is the approximate date given to the unification of Egypt and the start of her dynastic history. It is when the first cuneiform writing was developed in Mesopotamia and the first plough-shares used in China. In Europe, it is the beginning of Megalithic culture, when the first phase of Stonehenge was begun, and in South America it is when maize was first cultivated. We can therefore see this epoch as especially important for civilisations around the world, even if, as archaeologists assure us, they were isolated from one another. However, even though the people of the late Stone Age, in Europe as well as Egypt, were relatively knowledgeable concerning the stars and laid out temples such as Stonehenge to mark important dates on their calendar, there is still no evidence that they, any more than the people then living in Mexico, knew about the precessional cycle.

The other alternative, that the Olmecs and Maya inherited this knowledge from the Atlanteans, remains just a hypothesis. Assuming that the story of Atlantis is more than a myth and that the Atlanteans knew about precession, there is no evidence that this knowledge was retained by the Egyptians or anybody else. Although Edgar Cayce claimed that the Atlanteans constructed a hall in the Yucatán, he did not say that the secret cache it contained was accessed by the Mayans. Nor did he say that the Mayan civilisation contained anything more than traces of the much earlier Atlantean civilisation. Given that there is a gap of some 9,000 or 10,000 years between the supposed demise of Atlantis and the earliest long-count date yet found, it would be strange indeed if detailed knowledge of the precessional cycle were retained when so much else was forgotten. This being the case, we have to ask ourselves how else might the Maya and Olmecs have obtained the knowledge necessary to time their calendar to end at the solstice of 2012?

CIVILISATION FROM OUTER SPACE?

One possibility that we have so far not explored is extraterrestrialism. Might von Däniken have been right all along that the Olmecs, Maya or whoever else were visited by intelligent aliens from another star-system? Might it have been space-people who brought the knowledge and technology to calculate the shift in the equinox and the fact that in 2012 the winter solstice would occur with the Sun positioned at the star-gate?

In a small museum in Oaxaca, Mexico, I saw indirect evidence of this in the form of a most peculiar sculpture. It was of a head, but one that was not human. The head was divided in half, with one side of it almost humanoid except for his curious ears. The other half looked decidedly 'alien' and was reminiscent of something out of 'Dan Dare'. Looking at this sculpture, which did not have a caption, my immediate thought was that it was meant to represent a demigod, i.e. an individual who was part human and part alien in his ancestry (see Plate 22). Could this, I wondered, be the secret behind the advanced cultures of Central America: that their knowledge came from the stars?

The problem with this solution is that, while it provides a satisfactory answer to the central question of where the knowledge came from, it raises a host of difficult ancillary questions. To begin with, there is the technological question of how anyone, even intelligent aliens, could navigate the immensity of space. When even the closest star is 4.5 light years away (i.e. roughly 26,000,000,000,000 miles), how could an alien craft get here? Then, assuming this hurdle could be surmounted, there are problems of adaptation. Could a visiting alien survive the physical conditions and bacteriological lifeforms on our planet?

We need to remember that our planet, the Earth, exists in a precarious balance. Because we are so familiar with it, we don't give a second thought to just how unusual it is to have a world where much of the surface is covered by water. For this to happen, the planet has to be at just the right ambient temperature: too cold and all the water will freeze; too hot and it will boil away as vapour. Then there is the balance of gases in the atmosphere: roughly four-fifths nitrogen to one-fifth oxygen and a small percentage of other gases, such as carbon dioxide.

We, and the rest of the animals and plants on Earth, are closely adapted to its conditions. We are able to breathe the air, drink the water and, for the most part, cope with extremes of temperature between the poles and the equator. Even assuming a visiting alien came from a planet somewhat like the Earth, it seems extraordinarily unlikely that it would be exactly the same. Yet we know from our own experience that even small variations in, say, the amount of oxygen in the air or the salinity of the water have profound implications for life. That means our visiting aliens would almost certainly need to wear spacesuits at all times and would be entirely reliant on the life-support systems – food, drink, waste management and so on – that they had brought with them on their mother craft. These reservations apart, there may in my opinion be some substance of truth behind the notion of extraterrestrial contacts being made in ancient times, the key centre for this being Egypt.

The main reason why mainstream archaeology has set its face against diffusionism from Africa and Europe to America is because Mayan culture seems on the surface of it to have little in common with contemporary cultures elsewhere in the world. This, however, is to ignore certain similarities that lie just beneath the surface. The Mayan obsession with the bacabob, the gods of direction who hold up the sky, has much in common with ancient Egyptian beliefs concerning the sons of Horus. These gods of the horizon – *Duamutef* (jackal-headed), *Qebhsenuef* (falcon-headed), *Hapy* (ape-headed) and *Imsety* (human-headed) – are the prototypes of the symbols given by Christians to the four evangelists, Matthew, Mark, Luke and John, who are similarly linked to the four cardinal directions.

Egyptian parallels with Mayan beliefs went much further than this. In Egypt, the god Osiris, who died and was resurrected, was also a corn god. The mythology concerning his death and rebirth was as important to ancient Egyptians as the story of Jesus's death and rebirth is to Christians today. Indeed, the Osiris religion of Egypt was in many respects similar to Christianity. The legend of Osiris states that he, and a party of other anthropomorphic gods, were sent to the Earth by God to inaugurate civilisation. Osiris taught mankind how to domesticate animals and plants, observe the right methods of worship and instigate a proper legal system. His vizier, Thoth (called Hermes Trismegistus by the later Greeks), taught them mathematics, the sciences and the hieroglyphic system of writing. Under Osiris and his queen, his sister Isis, the country prospered and peace reigned. Unfortunately, they had a younger brother who envied Osiris his position. This brother, Sett, murdered the good Osiris and made himself king of Egypt. He scattered the parts of his brother's body along the Nile so that he should not have a proper burial. Thus was brought to an end the golden age of Osiris.

The rule of Sett was harsh and unjust, but Isis did not despair. She gathered together the pieces of her husband's body and then, using prayers learnt from the sun god Ra, brought Osiris back to life long enough to become pregnant by him. Osiris, his work on Earth done, departed for the stars. Isis, meanwhile, concealed her pregnancy and in due course gave birth to a son called Horus. He, on reaching manhood, challenged his uncle Sett for his father's crown. A series of battles was fought and in the end Horus was victorious. He re-established the rule of law and became Egypt's first true pharaoh after Osiris. Thereafter, all pharaohs, while they were alive, were looked upon as incarnations of Horus. After they died, they went through rituals (including mummification) that transformed them into Osiris. In theory, at least, they became one with the ascended Osiris in just the same way as Christians seek to become one with the ascended Christ.

We know now from the pyramid texts left behind by the ancient Egyptians that their funerary rituals were not only intended to transform the body of the pharaoh into something that looked like 'an Osiris' but to send his soul to their 'heaven': the *Duat*. This was a place similar to Earth where the righteous lived in peace and harmony under the rule of the benevolent god-king Osiris. However, unlike the Christian Heaven, which tends to be thought of as an invisible place occupied by spirits, the Duat was set among the stars: most specifically, the stars of the constellation Orion.

It seems that Orion held a special significance for the Egyptians because it stands to one side of the Milky Way. They looked upon the Milky Way, the 'winding waterway', as a cosmic counterpart of their own river: the Nile. Accordingly, during the IVth Dynasty, they built pyramids on the west bank of the Nile to represent the stars of Orion, those at Giza being symbolic of his Belt. Though this scheme seems to be incomplete, the intention is clear. By building a gigantic scale model of Heaven on Earth, they were establishing a psychic link to the stars of Orion/Osiris. Then, by carrying out certain funerary rituals (such as the 'Opening of the Mouth' and 'Weighing of the Heart') within the confines of the pyramids, they could assist the pharaoh's soul in its journey to the land of Osiris: the stars of Orion.

All this is covered in detail in *The Orion Mystery*, the book I co-authored with Robert Bauval. When it was first published (in 1994), it caused a furore. Though the book provided abundant evidence that the pyramids were intimately connected with the cult of Osiris and that this was essentially stellar in nature, it still attracted many critics loath to admit that the stars played such a big and crucial part in the religion of the ancient Egyptians. In part, this was an instinctive response to a theory that reminded them too much of Erich von Däniken's bestseller *Chariots of the Gods?*. However, there was also a more serious and scientific objection: where else other than the pyramids and in certain inscriptions and wall-paintings was there evidence that Orion played such a large role in the religion of the Egyptians, or anyone else come to that?

THE MAIZE GOD AND ORION

What we didn't know at the time was that even as we were putting together *The Orion Mystery* in 1993, on the other side of the Atlantic David Freidel, Linda Schele and Joy Parker were publishing their mould-breaking work *Maya Cosmos*.

Had we known this, it might have helped us greatly with our critics. For in this work, the authors, notably Schele, presented firm evidence that the Mayan maize god, just like the Egyptian corn god Osiris, was closely linked to a creation myth concerning the constellation of Orion.

The recent discovery of a Mayan wall-painting in northern Guatemala that features the maize god and is conservatively dated to *c.* 100 BC indicates that his story is even more ancient than was previously thought: indeed, it is contemporaneous with at least the latter days of the Olmec civilisation and is probably older still. Maize was regarded by the Maya as a gift from the gods: more specifically, the maize god. According to the *Popol Vuh*, they created our race of mankind out of maize dough, earlier attempts using mud and wood having failed. The maize god was therefore the prototype of mankind; he was our 'First Father'. There is no doubt that his death, decapitation, burial in the ball-court, impregnation of a maiden by spitting in her hand and finally resurrection by his victorious sons had symbolic meaning for the Maya and in some way mirrored the growth cycle of maize.

However, it is also clear that for them the maize god was more than the personification of the growth cycle of their staple food. We see this in the way that the image of King Pacal on the lid of his tomb shows him dressed as the maize god as he makes his way to the portal of heaven. He becomes First Father in just the same way as Egyptian pharaohs were thought to be transformed into the god Osiris by virtue of the rituals performed on their mummies after death. This is more than a strange coincidence, for Osiris was the equivalent of the maize god for the ancient Egyptians: he was the god believed to have been responsible for giving them the gift of corn.

This, however, is only one of the amazing parallels that connect the Mayan maize god with Osiris, for both are intimately linked with the stars of Orion. In the extended version of the rebirth legend of First Father, he was taken by canoe to Orion. This was the same place where, at the beginning of the age, the ancient gods laid three 'stones' (the stars Alnitak, Saiph and Rigel) to make a cosmic fireplace. These stars were in some sense thought of as counterparts to the three stones that made up the hearth in the middle of a typical Maya house. As such, they represented the concept of home as a place of warmth and their close connection with the sacred 'stuff' of fire.

According to Schele, the connection with Orion went even further than the mythical setting of stones by gods. She discovered that the Belt of Orion was associated with the symbol of a cosmic turtle, on the back of which were depicted three stars. She believed that a split in the back of this turtle, shown on a number of ceramic pots and plates, represented the ball-court where, according to the *Popol Vuh*, the ascended maize god had his residence.

Just like Osiris, the Father and corn god of the Egyptians, the First Father of the Maya ascended to the stars of Orion's Belt.[2] As *Maya Cosmos* was published in 1993, the year before *The Orion Mystery* appeared, it is unlikely that Schele knew how her discoveries concerning Orion and the rebirth myth

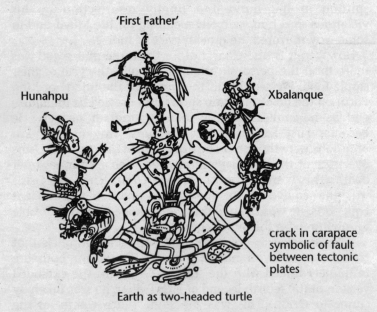

'First Father'

Hunahpu

Xbalanque

crack in carapace
symbolic of fault
between tectonic
plates

Earth as two-headed turtle

Figure 46: Resurrection of the maize god

of the maize god so exactly paralleled the Osiris religion of ancient Egypt. Though not stated, the implication was that our Earth (symbolised by a double-headed turtle) has a cosmic counterpart in the vicinity of Orion's Belt (symbolised by a turtle with three stars on its back). This planet – for that is what the symbol of the turtle indicates it to be – is evidently Earth-like. Now, this similarity in beliefs is extraordinary, especially when it is realised that Osiris was one of a family of gods who are said to have come to Earth to bring civilisation to human beings. The almost inescapable conclusion is that these 'gods', who the Egyptians believed had at one time ruled over Egypt, were anthropomorphic in form. One is therefore drawn to the conclusion that they were incarnated entities like ourselves, who came from the direction of Orion. I say this guardedly because I am well aware that the stars of Orion's Belt are distant from us by between 65 and 545 light years (LY). For us, these are impossibly large distances to contemplate navigating. However, we can perhaps assume that souls do not have to navigate space in any ordinary sense of the word. Maybe Orion really is the home from which First Father, Osiris and perhaps even other 'space-people' have come to bring the benefits of their civilisation to the Earth. Thus it was that, perhaps unwittingly, Schele and her colleagues produced much-sought-after evidence suggestive of extraterrestrial contacts between the Maya and 'gods' from Orion. The question remains, though: can this be true?

AVATARS OF THE MASTERS

The strange connection between the myth of the Mayan maize god, the Egyptian corn god Osiris and the constellation of Orion is not the end of the story. In 2003, a programme entitled *Stonehenge of the North* appeared on British television. The title was something of a misnomer, as it didn't concern anything that resembled what most people understand by the name Stonehenge, i.e. a Stone Age monument constructed from large, coarsely cut rocks arranged in a circle. In fact, the subject of the programme was a collection of three large, stoneless 'henge' monuments near Ripon in Yorkshire. These henges consisted of circular ditch-and-bank constructions, similar to the one surrounding Stonehenge, and each about 240 metres in diameter.

At ground-level, the henges seemed fairly unremarkable, but seen from the air they bore an uncanny resemblance to the photo of Orion's Belt published in *The Orion Mystery*. The striking similarity was immediately apparent and caused me to gasp when I first saw it. However, as the programme progressed it was revealed that, in *c*. 3000 BC, the Belt of Orion aligned with the entrances leading into the henges just as the star Sirius was seen to be rising in the east. This would be most striking, of course, on the day of Sirius's heliacal rising: its first appearance at dawn after its 70 days of invisibility. In Egypt, this was celebrated as the birthday of Horus, the son of the father Osiris (Orion), and it signified the start of a new year. The implication was that these henge monuments, and possibly many others to be seen throughout Britain, were monuments linked with a religious cult similar to that of the Egyptian Osiris/Orion. We know that there was a British cult of Orion in later times but this seems to indicate that it goes way back into the Stone Age.[3]

The approximate date when the Thornborough monuments were raised – 3000 BC – is also the date generally given for Stonehenge I: the original ditch-and-bank monument with its entrance aligned on the summer solstice sunrise. This preceded by centuries the building of the first stone circle in its middle. My computer revealed that, on the summer solstice at that time and seen from the latitude of Stonehenge, the Belt of Orion would have risen heliacally: at the same time as the Sun. In Egypt, the heliacal rising of Orion (rather than of Sirius) symbolised the rebirth of Osiris the corn god. Since even in the Stone Age southern Britain was very much given over to the cultivation of wheat and other cereal crops, it is not unlikely that, as in Egypt and Mexico, Orion was linked to the rebirth of a British corn god, his name lost in the mists of time. Thus the positioning of the henge monuments at Thornborough in Yorkshire as well as those in Wiltshire suggests this may have been a major cult throughout the country. This raises the interesting question: how is it that a cult of Orion features so prominently in the cultures of Egypt, Mexico and

Britain when, as far as we know, the people living in these countries were not in direct communication with one another? Leaving aside for a moment the possibility of maritime contacts, there is one other explanation as to how people living remotely from one another in time and place could have had similar ideas concerning Orion, and this is reincarnation.

Many people these days believe in reincarnation. In fact, taking the whole population of the world into account, there are probably more believers than non-believers. The widespread belief in reincarnation is not really surprising. The idea that the soul, while being a separate creation from the body, can only incarnate once and on the basis of its performance in that one life will be judged for all eternity is patently unfair. However, if, as reincarnationists believe, we live many times, then that evens the score. A just labourer could be elevated next time round to be born in the nobility; an unjust king could be condemned to live his next life as a beggar. In other words, our sufferings may not be quite as arbitrary as they seem but rather provide opportunities for self-development. This, to my way of thinking, makes much more sense of the inequalities of life than either the orthodox Christian belief in one life only or disbelief in the soul as an entity distinct from the body.

As we have seen, the doctrine of reincarnation was one of the most important themes to come out of the Edgar Cayce readings. In his trance condition, he stated categorically that many of his clients had lived earlier lives among the Mayans and earlier still as Egyptians. May not a memory brought over from one life to the other be the route by which common cultural ideas concerning the role of Orion as the maize god appeared among the Mayans some several thousand years after it appeared in ancient Egypt? This is an obvious possibility but is not, in my opinion, the most likely explanation. I believe we have to dig deeper into the Cayce readings and also look elsewhere to see what this might be.

Edgar Cayce had much more to say about the life of the soul after death than just its reincarnation potential. According to him, when people die their souls do not always stay within the region of the Earth but frequently journey to other planets or even stars. For example, he said that the person in reading 2454–3 (Eula Allen) had spent a period of time prior to her most recent incarnation in the environs of the star Arcturus: '. . . Arcturus comes in this entity's chart, or as a central force from which the entity came again into the Earth material sojourns. For, this is the way, the door out of this system. Yet purposefully did the entity return in this experience . . .'.[4]

In a question-and-answer session, Edgar Cayce revealed that he too had spent time in the Arcturus system before coming to Earth:

> (Q) The sixth problem concerns interplanetary and inter-system dwelling, between earthly lives. It was given through this source that the entity Edgar Cayce, after the experience as Uhjltd, went to the system of Arcturus, and then returned to Earth. Does this indicate a usual or an unusual step in soul evolution?
>
> (A) As indicated, or as has been indicated in other sources besides this as respecting this very problem, Arcturus is that which may be called the center of this universe, through which individuals pass and at which period there comes the choice of the individual as to whether it is to return to complete there – that is, in this planetary system, our sun, the Earth Sun and its planetary system – or to pass on to others. This was an unusual step, and yet a usual one.[5]

Edgar Cayce is not the only person to have claimed a special status for Arcturus, a 'red giant' star in the constellation of Boötes. In 2003, while on a lecture tour of South Africa, I was invited to have lunch with Dr. J.J. Hurtak, author of an extraordinary book called *The Keys of Enoch*®. I had first come across this book in 1994, shortly after the publication of *The Orion Mystery*, and I was keen to meet its author as I could see that he was, like myself, deeply involved in Hermeticism. I hoped to discuss the contents of his book, which, like the original 'Book of Enoch', was the record and commentary of a mystical experience.[6] *The Keys of Enoch*® is an epic work, comparable in some ways with William Blake's poems *Albion* and *Jerusalem*. I do not pretend to comprehend it in its entirety, any more than I do Gurdjieff's epic *All and Everything: Beelzebub's Tales to His Grandson*. However, picking through it I have found fragments which resonate with meaning and, for me at least, provide a certain inspiration.

Our meeting at the restaurant went well and we discovered that we held many ideas in common, even though, on an outward level, they were expressed in different ways and using different modes of thinking. It was clear from our conversation that he, like me, was deeply involved in the mysteries of Orion and its connection with incarnating 'Masters'. Meeting its author inspired me to take a fresh look at *The Keys of Enoch*® to see if there was anything else that might explain this connection further. What had struck me most in 1994 when I first came across it was that, unbeknown to Robert Bauval and myself, Hurtak had already linked the Giza pyramids with Orion and the cult of Osiris. Like us, he was convinced that the Pyramid Texts meant what they said when they linked Osiris with Orion. In a glossary at the end of the book, he wrote:

> Osiris: The Lord Creator from Orion that was responsible for one of the programming attempts by the Brotherhoods to raise the consciousness of the root races by showing the model of death and resurrection. Enoch [evidently Hurtak's guide during his own unique journey that he says took place in 1973] considers Osiris as Osi Osa, a twin deity of the Mid-Heavens, subservient to the Father Creator.[7]

These ideas were remarkably in harmony with my own discoveries, documented in my book *Signs in the Sky*, that the Orion constellation was, symbolically at least, the destination of the soul of more than one ascended 'Master', including Jesus Christ.[8] Hurtak went further than this. He claimed that Orion, along with its close neighbour the Pleiades star-group, was the true source of the spiritual power brought by Christ to mankind to enable their redemption: 'Kesil [Orion] emanates the Gnosis, the knowledge which creates the *Pneumatikoi*, the spiritual powers of Christ. However, the pre-physical garment of Light which is needed to embody this higher Light consciousness comes from the Pleiades . . .'.[9]

He says also that Orion contains a gateway or gateways into the 'Higher' universe through which redeemed souls must pass:

> The seventh key refers to the Metatron who receives us as we go through the gates of Orion from this Son Universe into the Father Universe of the Elohim Creators.
>
> And when I was taken to Orion, I beheld within Trapezium Orion, Light layers of great intensity which revealed a series of 'inner heavens' collectively forming the foundation for birth and regeneration. And I was shown how Trapezium Orion – the threshold gate of 'star creation' – is in conjunction with omega Orion, the region of 'star death'. Both are aligned with the Father's Throne governing through the star region of Alnitak, Alnilam and Mintaka.[10]

All of this could be dismissed as the ramblings of an eccentric mystic were it not that what is written in the *Keys* has a resonance both with ancient teachings and modern astronomy. The identification of Orion as a place of the birth and death of stars fits well with what astronomers are now discovering. Stars are being born inside the clouds of dust that make up the Orion nebula, itself the remnants of many dead stars. Also, within the Orion nebula, which lies in the 'sword' that dangles from Orion's belt, is a group of very large stars called the 'Trapezium'. This seems to be the region where Hurtak claims to have been taken during his journey with Enoch and this area is where astronomers believe they have found stars with rings of debris circling them that they believe are forming into planets. This was not known in 1973, at the time Hurtak was writing *The Keys of Enoch®*. The true significance of these star-fields in the Trapezium has only been appreciated since the launching of the Hubble telescope in 1990. In 1992, the possibility of proto-planetary discs in the Orion nebula was announced but it was a further two years before the Hubble telescope was in a fit state to reveal the discs. This discovery, some 20 years after *The Keys of Enoch®* was written, elevates the dust clouds of Orion from a curiosity to being one of the most exciting areas in the sky: a stellar nursery.

What is also very interesting is that, as Linda Schele discovered, the ancient Mayans linked the Orion stars Alnitak, Saiph and Rigel with the laying of a cosmic 'hearth' on the day of creation. The 'sword' region of Orion, where these star-systems are being formed, sits within the triangle formed by these three stars. It is the celestial 'fire' in the cosmic hearth. Presumably, it was from this hearth that they believed fire was brought to the people by their lineage god, Tohil/Kawil, at the start of our present age.

According to *The Keys of Enoch*®, we would be conscious of the relationship between the Earth and Orion had our planetary system not fallen under the control of certain 'Dark Lords'. Their place of origin is said to be in certain northern stars in Ursa Major (Big Dipper), Ursa Minor (Little Dipper) and Draco (Dragon). These entities, perhaps to be understood as the same as the 'Lords of Death' that the Mayans associated with Xibalba, came to Earth from these stars and corrupted mankind. The Dark Lords are evidently opposed by 'Lords of Light', who live in other star-systems – notably, of course, Orion and the Pleiades. Also, like Edgar Cayce's readings, *The Keys of Enoch*® gives special status to Arcturus, a giant red star in the constellation of Boötes, the Shepherd. Thus, Enoch's Key 201 is: 'The Key to the future astrophysics and cosmology is given in Archturus [*sic*] who is to be heard and understood as "one of the living sons of light".'[11] The commentary on this key contains the following:

> Our closest center for soul mapping is Archturus, the 'Shepherding Frequency of Light' which governs the preparation of Man for the coming of the Brotherhoods of Light; Archturus or 'Ash' means the 'Good Shepherd.' . . .
>
> Archturus is our Mid-Way station. It is the seat of our administration and is the thesaurus which holds the key documents used for governing the soul progression of our planetary intelligence.
>
> Archturus is also the first threshold of clearance for travel beyond our consciousness time zone.
>
> It is spoken by Enoch that we shall be taken up

through Archturus which is the clearing level of our
planetary creation through which we must pass to be
perfected before being programmed to go on to other
levels of creation.[12]

In his readings, Cayce claimed that while in the environs of
other planets or stars, souls are instructed in 'schools', whose
subjects of specialisation vary. On returning to Earth and
reincarnating in a body, they bring with them subconscious
knowledge and talents appropriate to the star/planet in
question. In this way, the music of a Mozart or the genius of
a Michelangelo could quite literally come from the stars.

Now to a rationalist, schooled in the teachings of Darwin
and in denial of the soul's very existence, let alone its
potential for reincarnating, such ideas must seem absurd.
Even for the more religiously inclined, these ideas pose
problems. 'Heaven' is a very vague concept for most
Christians, Jews, Muslims and even Buddhists. To suggest
that there might be many different gradations of
heaven/purgatory and these are associated with particular
stars and planets goes way beyond the teachings of the
Bible, the Koran or the Dhammapada. Yet it is curiously in
line with what we know about the ancient Egyptian and
Mayan religions. Perhaps we should accept that there is a
small possibility that they knew more about the afterlife
than we do. Maybe the idea that Lord Pacal's soul was
destined to journey to some star or planet in the vicinity of
the Belt of Orion was true after all. Von Däniken may have
taken things to extremes when he suggested that the Lid of
Palenque represented a spaceman at the controls of his
craft. However, if we read the images on the tomb lid as a
mythic narrative, then it does talk about a journey to the
stars. Perhaps Pacal's spirit did take leave of his body and fly
to 'the place of creation': the stars of Orion's Belt.

There are, of course, other stars much nearer to us than
those of Orion. In the general vicinity of Orion there is Sirius
(8.7 LY) and Procyon (11 LY). On the other side of the sky
there is the Eagle Star, Altair, which at 16 LY is slightly less
than twice the distance of Sirius and only four times the
distance of our nearest neighbour, Proxima Centauri (4.3
LY). All or any of these star-systems could have planets and

it is not beyond the bounds of possibility that at least one of them may also be home to a more advanced civilisation than our own. If star-souls from these systems have visited the Earth, then it is possible they too have seeded it with their ideas at different places and different times. If this is so, then we should perhaps not be surprised at similarities between cultures such as the Egyptian and the Mayan, even though they are separated by time as well as the Atlantic Ocean. It would also explain the Mayans' and Egyptians' apparent obsession with the stars of Orion.

The legacy which has come down to us from the ancient Maya and Egyptians in the form of myths and beliefs may be confusing. However, underneath the veneer of what for us are alien cultures, we can glimpse ideas and sciences that ought to have been beyond the capabilities of Stone Age peoples. It is hardly novel to suggest that at least some of the ancient 'gods' were spacemen from other worlds and it is easy to dismiss such ideas as crackpot. However, nearly all ancient cultures talk in terms of gods who are sufficiently anthropomorphic to associate with humans. If star-folk from a more advanced planet did visit Earth, then they would naturally be interested in establishing time and space coordinates for this planet. If they had computers at their disposal, then they could have discovered the precessional cycle and given the Maya a calendar that would reach its completion on 22 December 2012.

Unless we can disprove the logic of Einstein's physics, to drag a 10,000-ton 'Starship *Enterprise*' across the vastness of space, at warp-speeds approaching the speed of light, is clearly impossible. Thought, however, travels in an instant. With this new vision of soul-travel in the afterlife, the universe suddenly becomes much smaller. Who is to say, then, that the soul, riding in a vehicle of light – referred to in the *Keys* as a '*merkabah*' – is unable to cross the great divides that separate one star-system from another? What is more, if we accept the possibility that after death our souls can travel through space and visit the stars, then it follows that it must also be possible for what we might term 'alien' souls to visit the environs of planet Earth.

Now, if we allow for all this as a possibility, one which I admit goes far beyond the scope of physics, then the religion of the Maya begins to make more sense. We can also see how they may have come upon their long-count calendar. Perhaps one or more 'masters', comparable with Osiris in Egypt or Krishna in India, incarnated among them and taught them great wisdom. The conjuration of ancestral spirits at times of zenith culmination of the Sun, Moon and planets could also have extended to communicating with extraterrestrial spirits: in their parlance, the gods who taught humanity how to cultivate corn. Also, reading again the story of the hero twins, it seems to me that for them Xibalba was not really a subterranean Hades. It was the earthly plane where we live, ever in fear of death but hopeful of an afterlife. Perhaps the hero twins, like Osiris and the other anthropomorphic gods of Egypt, are to be understood as visiting souls from some advanced civilisation that exists on a planet somewhere in the region of Orion. The 'canoe' in which Linda Schele says the maize god is transported back to the stars of Orion doesn't have to have been a metallic ship like a space shuttle. If Hurtak is correct in what he reports in his *Keys*, it could have been something much less tangible: a merkabah.

In any event, the arrival of incarnating masters with knowledge and powers would have seemed inexplicable. It seems that what First Father and the hero twins taught was the immortality of the soul and therefore the pointlessness of fearing death. By demonstrating their own ability to defeat death by reincarnating, they effectively neutered the Lords of Fear, who at the time held the Mesoamerican world in thrall. Their task achieved, they departed back to the stars from which they came.

Now this might seem a very odd way of reading the *Popol Vuh* but in reality it is no stranger than what most Christians believe about Jesus Christ. Read without prejudice, the story of the New Testament is similarly one concerning an avatar of the godhead who comes to Earth. Jesus teaches compassion to his followers, is sacrificed, resurrects himself and then ascends to heaven. Curiously enough, as readers of my earlier book *Signs in the Sky* will be aware, on the day of Jesus's Ascension into heaven (which I have shown to be 27 May AD 29), the Sun was positioned exactly at the northern star-gate over the upstretched hand of Orion. If this is a coincidence, then it is certainly a very strange one.

Readers of *Signs* will also be aware that the prophecies of Jesus (given in Chapter 24 of the Matthew Gospel) indicate that the star-gate over the hand of Orion gives the timing for the end of the current age. Is it therefore a coincidence too that the Mayan prophecy for the end of the age is also linked with a star-gate, in this case the one in Sagittarius? I think not. At the very least, it behoves us to look again at what is happening in our world, as we apparently stand on the brink of what are prophesied as Earth-changes of barely imaginable proportions.

Taking all of this evidence together, we can see that the purpose of the long-count Mayan calendar was to point to a marker-date: a time when the Sun aligns with the southern star-gate exactly at the winter solstice. This date, so clearly defined by the calendar, has all the hallmarks of an appointment with destiny. Whether that destiny is simply an Earth in upheaval or a brief time of chaos before the emergence of a new world order is not clear. However, if we accept that the calendar was given to us by intelligent spirits who visited Earth from outer space, then 22 December AD 2012 could be when they plan to return. That makes it a date with destiny that we should ring in our diaries.

CHAPTER 12

Cycles of the Sun

In early June 2005, I was contacted by Mitch Battros, a man who has made it his personal mission to alert the world to the effect the Sun is having on climate change. Mitch, who in 1995 set up his own broadcasting station called 'Earth Changes TV', wanted to know if I was interested in coming on a cruise to Mexico and Belize. While there, I would witness a Mayan fire ceremony and hopefully gain some further insight into the workings of the Maya today. In return, I would give a few lectures to his other guests – a party of around 30 – and explain to them the importance of the Mayan calendar. Naturally, I said yes to this unexpected invitation – partly because it sounded like fun but also because I wanted to meet Mitch himself. Although I had been interviewed by him several times before, this had been over the phone, with him in America and me in Britain.

Mitch's stated aim in setting up his TV/radio station was to form a bridge between hard-facts science and what he terms 'Foo-Foo': irrationally held but possibly true beliefs held by millions of searchers after the truth. While the balance of his work steers towards the scientific, he has not been afraid to invite onto his show people such as Mayan elders and other proponents of ancient wisdom. Among his regular guests are Carl Johan Callemann (author of *The*

Mayan Calendar); Dr Tom Van Flandern (for 20 years, director of the US Naval Observatory, and cofounder of Meta Research Inc., a scientific, non-profit organisation dedicated to investigating currently unfashionable aspects of astronomy); and Dr Paul La Violette (author of *Genesis of the Cosmos* and *Earth under Fire*, which concern the galactic causes of Earth-changing events that are barely remembered by the human race but form the basis of many myths). Other guests have been the volcanologist John Search; Adam Rubel (cofounder of *Sak-Be'*, an institution for the study of Mayan and other indigenous spiritual traditions); Professor Ed Mercurio (famous for his work on the effects of galactic cosmic rays on climate); Dr Casey Lisse (from the Department of Astronomy of the University of Maryland and closely involved in the 'Deep Impact' project that brought about the first rendezvous with a comet); and Larry Combs (lead forecaster for NASA/NOAA and an expert on sunspot flares and mass ejections).

Given the high calibre of so many of Mitch's guests, I feel honoured that I have also, several times, been invited onto his show. The first time was a transatlantic telephone discussion relating to *The Mayan Prophecies*. At that time, the book had only recently been published, and I was pleased to discover that Mitch was on board regarding its major revelation: that the Sun goes through cycles of activity that have profound consequences for life on Earth. Since then, Mitch has himself become something of an expert on solar phenomena (such as sunspots) and the profound effect that these have on our weather. This he has distilled into an equation which he first published in 1997. Put simply, this goes: sunspots => solar flares => magnetic field shift => shifting oceans and jetstream currents => extreme weather and human disruption. It is an equation that is deceptively simple in its expression but has potentially devastating implications – as the residents of New Orleans will now testify.

Following his invitation to go on the Maya cruise, I met up with Mitch at a hotel in Houston on 11 June 2005. The following day, we proceeded to Galveston, where our cruise-liner, *Elation*, was docked. The group travelling with us numbered 30, mainly Americans but also a few Canadians and Britons. In high spirits, we set sail, each person hoping for some small illumination from the experience. The Gulf of Mexico was as calm as a mill pond, giving no indication of how, in just a few weeks' time, it would be whipped up into a frenzy by Hurricane Katrina. The only clue to what was to come when the hurricane season got under way was the extraordinarily high temperatures. Even out at sea, with a steady breeze created by the movement of the ship, it was too hot for me to stay for long up on deck. I was not surprised to hear that the sea temperature was already unusually high for the time of year. Experienced members of the party wagged their heads sagely and warned that this portended hurricanes in the months to come. How right they were was confirmed when Katrina and several other major storms wreaked havoc on not just Louisiana but also Cuba, Mexico and Florida. This, however, all lay in the future. For the present, I looked forward to a relaxing voyage and the opportunity to share ideas with like-minded people.

Our first port of call was Progresso, some 20 miles from Merida and on the top of the Yucatán Peninsula. I was eagerly anticipating this stopover, for though we would not have much time on shore, I was keen to pay a return visit to the ruins of Dzibilchaltun, an early Classic Mayan city whose name means 'Place with writings on the stones'. Most of the party were planning on going to Chichen Itza but a group of five or six decided to join me and we hired a small bus for the day.

Dzibilchaltun was everything I remembered it to be: a fine collection of Mayan temples mostly set on a long avenue leading towards a cenote, or watering hole. What I was primarily interested in, though, was not so much the buildings themselves as the orientation of the roads leading to them. On one such road, running north–south, there were set at intervals several tall flat stones. If there had ever been any writing on these stones, it had long since weathered away. However, they must have been important as each was located on its own little plinth. They looked like memorials and I presumed that each stone was erected at the termination of a katun in the same way that the highly decorated stones of Tikal were.

Leaving the stones, I walked down the kilometre-long road leading to Dzibilchaltun's most famous monument: the 'House of Dolls'. As it was around noon, the Sun was beating down ferociously, rendering every step an effort. By the time I reached the building itself, my shirt was drenched in sweat and my water bottle almost empty. It contrasted greatly with my memory of the same building.

The last time I had visited it was at dawn a day or so after the spring equinox. On that occasion, it was raining and there was a positive chill in the air. This had been a great disappointment as the major attraction of this building is that, uniquely among Mayan buildings, it is orientated in such a way that the rising Sun at the equinox shines directly through its open doorway and windows. Because of the clouds and rain, I had not been able to see this for myself and neither would it be possible now as the time and date were wrong. I did, however, check its orientation using my GPS. This confirmed that it was exactly on the east–west axis and therefore would indeed be aligned with the rising Sun at the equinox.

I climbed up inside the building to view the road leading to it from above. This produced another surprise. It was clear that the building itself – named House of Dolls on account of a collection of small, doll-like sculptures found buried there – was built inside a large Canamayte square. Furthermore, a little way back up the road leading into the main part of the city complex was another standing stone and this too was set upon a Canamayte plinth. As the site is less than 20 miles from Merida, I felt sure it had been visited many times in the past by Don José Diaz Bolio and that he must have been excited when he first saw this. Sad to say, it was not possible on this occasion to pay him a visit as he had been dead for some eight years by then. It was, however, reassuring to see further evidence of how important to the ancient Maya was the symbol he had rediscovered when examining the pattern on the backs of his rattlesnakes.

Leaving Dzibilchaltun, we returned to our ship and continued the cruise. Our next stop was the island of Cozumel, from where we were able to take a boat to the attractive resort of Playa del Carmen and then a coach to the Tulum. Again, the heat was unbearable but it was worth putting up with it to see these unique Mayan ruins, situated as they are right on the coast. Like Palenque, this was one of the cities visited by Stephens and Catherwood during their extended itinerary of the 1840s. In his book, Stephens complains greatly about the mosquitoes they encountered when encamped for the night in the 'Castillo'.

One useful consequence of visiting during the heat of the day was that the mosquitoes were seemingly at bay. What struck me most, though, was not the buildings, nor even their idyllic setting next to a beach, but rather the fact that enclosing the city was a high and very well-built wall made from stone blocks. At first, I thought this was of modern construction but on rereading Stephens's book I discovered it was there when they visited. It seemed to confirm what they suspected: the city was occupied by its native inhabitants long after the Spanish conquest of the rest of Yucatán. Since no other Mayan city that I have come across has any sort of perimeter wall like this, it suggested that they learned the

Lithograph by Frederick Catherwood of Castillo at Tulum

Figure 47: Ruins of Tulum

idea from the Spanish. Unfortunately for them, even a wall of stone was no protection from cannon-fire, and still less from the predations of smallpox. By the time Stephens and Catherwood visited, Tulum, like almost every other city of the Maya, was an abandoned ruin, silent but for the ghosts speaking through time from its inscribed walls.

Our last port of call was Belize, capital of the country now of that name but which during my childhood was still known as British Honduras. That alone led me to expect differences with the neighbouring province of Quintana Roo, which is part of Mexico. The fact that Belize was once a British colony and is now a member of the Commonwealth would mean, I suspected, that its inhabitants, whatever tribal language they used among themselves, would speak English rather than Spanish as their international tongue. What I hadn't expected but what was very obvious from the moment of our landing was just how Caribbean this country is. For, unlike in the Mexican part of the Yucatán Peninsula, where the predominant ethnicity is Mayan, the larger part of the population of Belize is Afro-Caribbean in appearance: a legacy of slavery in the British Empire from the eighteenth and early nineteenth centuries. I had no time to investigate Belize City itself, which seemed a rather run-down outpost of colonialism. Instead, I had to hurry to catch a coach which was to take us to another destination, the ancient Mayan city of Cahal Pech. To get there was a long drive through first the rather squalid suburbs of Belize itself and then into the jungle areas close to the border with Guatemala.

A FIRE RITUAL AND INVOCATION OF THE JAGUAR GODS

Travelling with us was Mayan Elder Carlos Barrios, a gentleman, as his name would suggest, of Spanish blood and appearance. Appearances, however, can be deceptive. Carlos was born in and grew up in the highlands of Guatemala, his home being Huehuetenango. This lies far away from Guatemala City and many of the people of the area are of Mayan descent, most of them belonging to the Mam tribe. They have preserved much of the old ways and Carlos's early contact with the Maya led him into anthropology and eventually to his own initiation as a Mayan *Ajq'il*, or ceremonial priest and spiritual guide, with the Eagle clan.

After a brief speech by Carlos, given in the visitor centre by the gate, we went into the precincts of the ancient city. Most of the site turned out to be overgrown, the pyramids buried in the earth with trees growing from them in a way not dissimilar to prints of Palenque made by Frederick Catherwood in the 1840s. Unfortunately, we had little time to investigate the site before getting on with the fire ceremony itself. What was clear, though, from even a very cursory look, was that this must have been a major city during Classic times. After I returned home, further investigation revealed that not only was it of considerable importance at this time but it was probably founded in *c.* 1000 BC or even earlier. This would put it back into the era of the Olmecs, when the people of La Venta were busy sculpting heads out of great boulders of basalt.

Carlos leading the way, we walked through one plaza into the next. We found ourselves inside a large square which at one time must have been used for religious rituals. To the north was a very steep-sided pyramid that looked not dissimilar from those at Tikal rather than Palenque. The other sides of the square were bounded by walls and stairs rather like an amphitheatre. I could well imagine that in the past people would have gathered here to watch ceremonies performed either in the square itself or on top of the pyramid. It therefore seemed to be appropriate that Carlos was planning on performing the fire ceremony here, in the midst of this ancient sacred space, though I was still rather surprised that he had succeeded in gaining permission.

While the rest of us were spending a few minutes examining some of the buildings of the first court, Carlos changed into his ceremonial garb. This consisted of a white open-necked shirt that hung outside his jeans, an elaborate and highly coloured shawl, a crimson scarf worn over the head like a turban and, most impressive of all, a cummerbund-like belt made out of red material with its end hanging down in front of his trousers. This was very carefully tied and was exactly like similar belts worn by kings Pacal and Chan-Bahlum on the entablatures of the Cross Group of Temples at Palenque.

We gathered around Carlos in a circle and he began by drawing a 'sand' diagram on the ground using maize-flour as the medium. This consisted of a circle, roughly 4 ft in diameter, which he then divided into quadrants with two lines, perpendicular to each other, terminating in arrows. Then, in the centre of each quadrant, he drew a small circle and a fifth circle, of similar size to the others, around the centre of the diagram. Moving to the east side of the large circle, he drew a strange hieroglyph containing what looked like the profile of a flat-backed stepped pyramid. Next to this, he drew two dots, a symbol like a crescent moon and one or two other hard-to-interpret figures that looked like snake's fangs. Going over to his bag, he took out five balls of copal incense and placed one on each of the inner circles. The final drawing looked like those in Figure 48.

The initial drawing complete, he stood up and delivered a short speech, the gist of which is as follows: 'This is a glyph that has been used for 1,000 years. When this humanity survived its last destruction, there came from the Pleiades the four *balonev*, or jaguars, and they were charged to teach humanity. At this moment, we were in the last period of the glaciation. People were surviving all over Mother Earth by living in caves. They asked the jaguars, who were demigods, to clean the space-environment so that the sunlight could enter. They returned to Mother Earth and made a path for seven sacred altars on mountains. Then finally arrived one of the most important healers, that we call *Akabal*, meaning Venus, the new light, and then they called the Father Sun. Each of the jaguars arrived carrying incense for making the ceremony and each took one of the corners. That's the meaning. The circle is Mother Earth and the arrows are because the four corners point in the four directions of the universe. The central circle is this place, where we celebrate now. The four circles are the four corners and also the four elements: the fire, the earth, the air and the water. The central circle is the ether element. At the same time, the four circles symbolise the eyes of the Great Spirit in front of our eyes. That means we have a connection through the ceremony between ourselves and the Great Spirit. We don't need an intermediary. We have our own connection with the Great Spirit.

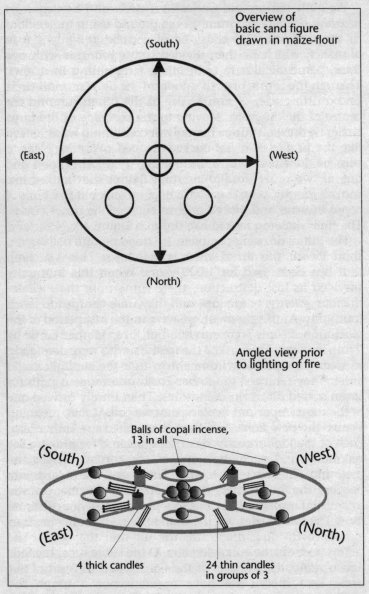

Overview of basic sand figure drawn in maize-flour

(South)

(East)

(West)

(North)

Angled view prior to lighting of fire

Balls of copal incense, 13 in all

(South)

(West)

(East)

(North)

4 thick candles

24 thin candles in groups of 3

Figure 48: Carlos Barrios's fire diagram

'And the Father Sun arrived here on the Earth and he talked to the jaguars and he said, "It's OK that you are creating life and we have the responsibility for this new humanity. Any time that you need something, you can call on me. But don't only call me when you have problems. Remember me also when you are happy and you have everything: when you have health, when you are prosperous and happy. Because that is the purpose of life: to be happy." This is our own responsibility but we forget to say thank you to the Great Spirit, the Great God. It's like I explained before [during his speech given in the visitor centre]: we each live for our own purposes. But we are destroying our Mother Earth. We are destroying our great house of Mother Earth. The purpose of the ceremony is to ask for our personal purposes: for fortune, health and economic stability. But essentially the focus of what we are going to ask is for the forgiveness of Mother Earth, for help so that we can create harmony and balance between the energies. That's our purpose.

'Today the day is date of *Sikim* [sic] who is the fortune, who is the intermediary between us and the Great Spirit. He is the eagle, the condor, the quetzal, the bird of power and is fortunate.[1] We can ask today for fortune that we have enough resources for ourselves that are essential for self-development. That's what I would like all of you to focus on but you can ask what you like.'

With that, he began laying out more copal balls, placing a further four around the central one at the middle of the symbol. He then broke open a packet of candles and, standing aligned in turn with each of the arrows, placed these in groups of three around the figure. He placed four thicker candles that were red in colour on the arrows, maintaining all the time the symmetry of the figure. As he did so, the noise of cicadas rose in volume until it was hard to hear anything else.

This part of the work complete, Carlos lit several candles that he held in his hand. Then, beginning at the east of the circle, he recited prayers of invocation to the first of the Jaguars: Bolon Quitzé.

'We call on you, Bolon Quitzé. You are the guardian of the fire of the red race. You are the guardian of the light. You are the energy of the spirit. We are calling you because my brothers come here and sacrifice. Please see our purpose and forgive us. We are asking that the light be spread around the Earth. I call you, Bolon Quitzé, because you are the energy of the light. You are the energy of the creator. You are the beginning and we are calling on you to give us the power that we are your warriors. On this date of *Quel Sikim*, we call the energy of the spirit of Sikim to help us with all our prayers, personal and communitary, prayers for Mother Earth. You can make possible the Great Spirit to turn his eyes and pay attention to the small ceremony. But you don't see the quantity, you see the force, the energy and the belt of power inside of the spirit of the heart of all my brothers and sisters. I hope you receive this energy. In essence, Bolon Quitzé, we ask for the light to guide humanity so that we can help Mother Earth to make the transition to the five suns and five elements. Bolon Quitzé, thank you.'

With that, he turned and faced the opposite direction, this time invoking the second of the jaguar gods: *Bolon Acal*. This god was evidently the guide and guardian of the black race in just the way that Bolon Quitzé was linked with the red race. He was invoked as the jaguar of the night, the light in the dark but one who puts obstacles in the path, who challenges the light. Then, facing south, Carlos invoked *Mahucutah Bolon*, guide and guardian of the white race, and uttered similar prayers. Finally, turning towards the north, he called on *Iquibalam*, the guide and guardian of the yellow race.

With these opening rituals complete, we moved on to the main event. Carlos distributed candles and cigars to every member of the group. We placed both the candles and the cigars within the outer circle, breaking up the latter to make material for the fire. Then, after spitting what seemed to be neat whisky onto the now heaped up pile of candles and tobacco leaves, he set it alight. Very quickly, the fire took hold so that, notwithstanding the earlier rain, within minutes the entire circle was filled with flames.

We now moved into a more personal phase of the ceremony, when Carlos went round the circle one by one. Facing each person in turn, he carried out a healing ceremony on them, each time placing on their head and in their right hand a small bag. Made of canvas and red in colour, it was about 2.5 in. in diameter and 8 in. high.[2] With each person, he uttered prayers and brushed his hands down from their heads to their shoulders and their upper arms. With some individuals, who he evidently saw were sick and needed something else, he would do other variations on this procedure. Then, the healing of individuals accomplished, he began a more generalised healing ceremony for the Earth. Most of this I could not understand, except that it again involved invoking the jaguars and each time running through a litany of all the day-names and numbers. At this point, the ceremony had to be cut short as we needed to get back to Belize City in time to catch the boat. Failure to do so would have meant it sailing without us and our having to make our own arrangements to get back to Houston. Reluctantly, but filled now with the good vibrations of the Maya ceremony we had witnessed, we reboarded the coach and headed back.

Unfortunately, as Carlos was travelling separately from most of the group, it wasn't possible for me to put to him all the questions I wanted to ask. However, what had immediately struck me about the ceremony was how similar in so many respects it was to those described by Landa and others as still being widely celebrated at the time of the conquest. The division of the circle into four with a central *quincunx*, or fifth point, equates exactly to the Bacab rituals and also those associated with invocations of the chac rain gods. The names of the jaguars are the same as those given in the *Popol Vuh*, where they are described as the fathers of the human race. The gods create them but find they are too perfect and know too much, so they spoil them just a little, weakening their eyesight and therefore their understanding of creation. Wives are then created for these four jaguars and from these semi-divine couplings are born all the tribes of the human race.

What particularly interested me was Carlos's speech at the beginning of the ceremony when he said the jaguars had come from the Pleiades to help the Earth at the start of the present age. In the *Popol Vuh*, these jaguars are lineage gods of the various tribes such as the Cauecs and Quiché. No mention is made of the Pleiades or of the jaguars being associated with the red, black, white and yellow races. However, this is a logical extension. It also ties in neatly with the Mayan wheel of colours: red–east, white–north, black–west and yellow–south. In Landa's account, the four gods linked with the directions are the bacabob. However, as was noted by J. Eric Thompson, almost everything and everyone has the potential to be divided by type according to compass directions, so why not the jaguars? Their connection with the Pleiades is not stated in the *Popol Vuh* but it should be clear that they are linked with those other 'fathers' of the human race: the two pairs of hero twins. This is clear from the fact that Hunahpu and Xbalanque are always shown with jaguar spots on their bodies. They are the jaguar people as well as being the maize gods. In other traditions, they may be thought of as coming from Orion, but that constellation is in the same part of space as the Pleiades and, if J.J. Hurtak is to be believed, entities from both places worked closely together to create elements of our genetic code that make us different from the animals.

I did not have time to discuss any of this directly with Carlos, but after returning home I consulted the Internet to see what else he might have said concerning the end-date of 2012. Here I found an intriguing article by him entitled 'The world will not end'.[3] In the article, he castigated the idea that the world is going to end in December 2012. He said the Mayan elders are very angry that people have been saying this. What is going to happen, he said, is a transformation.

The article then goes on to trace the connection between what happened when the Spanish arrived in Mexico in 1519 and the 'Harmonic Convergence' of 1987. It explained that the Maya had no fewer than 17 calendars. The one that they give most attention to today is the *Tzolk'in* or *Cholq'ij*. This, as we have seen, is 260 days long and is generated by counting the numbers 1–13 against the 20 basic day-names. What I didn't know and the Mayan elders were now revealing was that this count is also based on a cycle of the Pleiades. According to the elders, a cycle of 9 periods of 52 years (Aztec 'sheafs') began on Easter Sunday, 21 April 1519. This was the day when Hernán Cortés arrived in Mexico. He fulfilled a prophecy that an important ancestor (Quetzalcoatl?) would return 'coming like a butterfly'. The article says that what this was referring to (though the original prophet would not have known this) was the way the flapping of the sails on the Spanish galleons resembled the beating of butterfly wings.

The arrival of Cortés, though seen by Europeans as bringing civilisation to Central America, for the natives heralded the era of the 'Nine Bolomtikus' (hells), each lasting for fifty-two years. During this time, they were to have their land, freedom and self-respect taken from them. This period ended on the day of 'Harmonic Convergence': 16 August 1987. This, the elders said, was celebrated by many people worldwide as they prayed for a peaceful transition into the coming era of the Fifth Sun.

Reading what Carlos had to say did indeed bring back memories of Harmonic Convergence, a day when I and many of my friends gathered together in Dorset and offered prayers for world peace. I had not realised, though, that this was connected with the ending of the period of 'Nine Hells' that was initiated by the arrival of Cortés at Santa Cruz. I remembered, though, that the chief proponent of Harmonic Convergence, at least in the West, was José Arguelles, a former professor of art from Boulder, Colorado. As I had his principal work, *The Mayan Factor*, in my library, I decided to take another look at this to see what he said about the connection between the Mayan calendars and Harmonic Convergence. I found the following:

> The Mayan Matrix, the Tzolkin or Harmonic Module, bearing the code of the galactic harmonic, informs all systems with a common regulatory resonance called the light body. Just as each living organism possesses a light body – the DNA infrastructure – and even the entire species has its common collective light body, so the planet, as a conscious organism, is also characterised by its evolving light body.
>
> Like the light body of individual and collective organisms, the planetary light body is the consciously articulated resonant structure that regulates and allows for the fulfilment of evolutionary destiny. It is important to bear in mind that the planetary light body, embedded in the planet memory program, can

only be activated by conscious, cooperative effort. As we shall see, the key to the conscious articulation of the planet light body is in the science generically known as geomancy – Earth acupuncture.

As the radiant information bank of the planetary program, the 260-unit galactic code can be envisioned as primordially imprinting the electromagnetic ether of the *outer* planetary sheath, the upper of the two radiation belts that girdle the Earth. I say primordially, because the galactic core, Hunab Ku, like a powerful radio station, is endlessly generating the radiant light code.

The information flow between a planet body like Earth and the galactic core is maintained and mediated by the solar activity known as the binary sunspots. Both Sun and planet operate with the same galactic information bank. Whenever a small stellar body, such as our Sun, begins its evolutionary course, it is imprinted with the 26-unit galactic code. Once a planet such as ours attains a point of resonant activation, the galactic information flow mediated through the sunspots imprints the outer electromagnetic sheath with the basics of the planetary memory program.[4]

What the author seems to be saying here is that there is a link between the 260-day Tzolkin cycle – in many respects the most important of the Mayan calendars – and the larger, galactic cycle of roughly 26,000 years that we know from the precession of the equinoxes. The core of the galaxy, which is now thought to be a black hole (and was known as *Hunab Ku*[5] by the Maya), emits coded information in the form of electromagnetic radiation. This, he claims, is received by the Sun and relayed to the Earth via sunspot activity. Now, this was very interesting to me, as one of the major themes of *The Mayan Prophecies* was my co-author Maurice Cotterell's theory concerning long-period sunspot cycles and their impact on the rise and fall of civilisations. It prompted me to take another look at this strange phenomenon.

MAGNETIC STORMS ON THE FACE OF THE SUN GOD

At the time Cotterell and I were writing *The Mayan Prophecies*, the study of the impact that the Sun has on our weather was in its infancy. Since then, modern science has caught up considerably. It has revealed something our ancestors guessed intuitively: that the weather on Earth is governed by the 'mood' of our local 'god', the Sun. There is indeed a link, other than light and heat, between the Sun and the Earth, and this link is magnetism. The Sun, like the Earth and some (but not all) of the other planets, has a magnetic field.

It was once thought that the Earth's magnetic field existed because it had at its core a large and solid lump of magnetised iron. We now know that the situation is much more complicated than this simple model would suggest. Our magnetic north pole is not coincident with the geographic pole but is offset by some 600 miles. Furthermore, it is on the move. It is currently located in northern Canada, but, drifting as it is at a speed of roughly 25 miles per year, it could soon be in Siberia. This is not all. It is also clear from the geological record that the polarity of the Earth's field – which pole has 'north seeking' properties and which 'south seeking' – goes through periodic reversals. In layman's terms, north becomes south and south becomes north. This is not consistent with the idea of the Earth's core being a permanent magnet. It suggests that it is more like an electromagnet: one whose field is not stable but is induced by a flow of electricity through the core material.

The Earth's magnetic field is not confined to its surface but reaches far out into space. It forms a magnetic envelope around the Earth called the magnetosphere: what José Arguelles refers to as its 'sheath'. Though it is invisible and we are not aware of its presence, the magnetosphere is extremely important to life on Earth. For one thing, it protects us from damaging radiation produced by energetic particles hitting the upper atmosphere. Contained within the magnetosphere are the Van Allen Belts. Protons and other positively charged particles are trapped in the lower zone of these at about 3,000 miles above the Earth's surface. Electrons are trapped higher, in the upper zone, at around 10,000 miles of altitude. These particles, which would otherwise impact straight into the Earth, spiral backwards and forwards along the magnetic flux lines of the magnetosphere. They are eventually drawn towards the poles, where they produce the fantastic displays we know as the aurora borealis, or Northern Lights.

What was not known until recently was that the particles causing this phenomenon are of mostly solar origin. Four hundred years ago, Galileo trained his telescope on the Sun and revealed, much to the annoyance of the church, that its smooth face was pockmarked with 'sunspots'. At the time, the cause of these spots was not understood but the discovery that God's beautiful Sun had such blemishes was enough to send the Vatican into apoplexy. For this, and that other crime of science – saying the Earth goes round the Sun and not vice versa – Galileo was imprisoned. However, this did not deter others from studying sunspots in detail. It was subsequently discovered that the numbers and positions of spots (closer or further away from the Sun's poles) went through a cycle of roughly 11.1 years. A theory to explain this apparently irrational behaviour did not come until

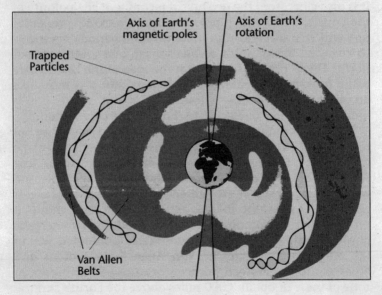

Figure 49: The Earth's magnetosphere

1961, when George Babcock proposed that the sunspots are caused by the differential rotation of the Sun's equator relative to its poles, with consequent effects on its magnetic field.

The reasoning behind this theory is quite straightforward. While the Earth's magnetic field is a simple dipole, with 'north' and 'south' poles near to its geographic poles, the Sun has a more complex field. In addition to a polar 'dipole' field, it has a 'quadripole' field around its equatorial region.

Further observation indicates that, unlike the Earth (which, being a solid body, rotates uniformly), the Sun exhibits differential rotation. Thus, while its poles rotate once every 37 days, its equator moves much more quickly, making a full rotation in only 26 days. The effect of this differential rotation, coupled with the Sun's complex magnetic field, means that lines of magnetic flux get wound up like spaghetti on a fork. This causes, deep inside the body of the Sun, regions of intense local magnetism. When this magnetic flux reaches a critical level, magnetic loops pop out from the Sun's surface. This has other effects, for where these loops protrude, the surface temperature of the Sun is considerably cooler than elsewhere: perhaps 4,000°C instead of 6,000°C. These areas of relative coolness (and therefore reduced luminance) can be up to fifty thousand miles across, which is roughly five times the diameter of the Earth

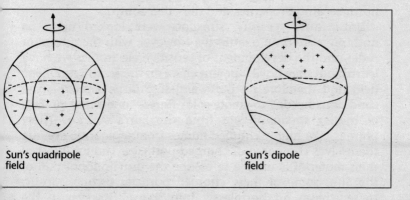

Sun's quadripole field

Sun's dipole field

Figure 50: The Sun's magnetic field

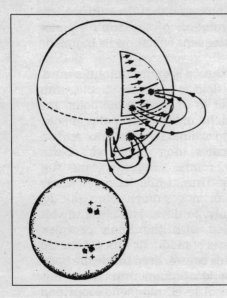

Sunspots are formed in pairs and are the result of magnetic looping. Where the loops erupt from the surface of the sun, the temperature is cooler and looks relatively darker than the rest of the sun's surface.

Figure 51: Formation of sunspots

and two-thirds the diameter of Jupiter. Because they are cooler and less luminescent than the rest of the Sun's surface, we see them as sunspots. Being connected with the magnetic loops, spots always appear in pairs: one with north-seeking properties and the other south.

Until fairly recently, sunspots were looked upon as nothing more than a curiosity. However, with the advent of radio and the development of sensitive electronics we have learnt that they have a potent effect on the Earth's magnetic field and therefore on our weather. Sunspots provide the conditions needed to create solar flares – enormous plumes of matter streaming out from the Sun's surface. These plumes, like highly charged flames from a gigantic fire, are themselves composed of charged particles. Though most of their material is unable to escape the gravitational pull of the Sun, some of these charged particles escape. Mostly, these stream off harmlessly into space. However, if the sunspot from which a plume emerges is aligned with the Earth, then charged particles hit the Earth's magnetosphere a few days after the appearance of the plume. They then ricochet backwards and forwards, following lines of magnetism running between the Earth's poles, before eventually being drawn into the lower atmosphere. As they do so, they produce the spectacular firework display known in the north as the aurora borealis and in the south as the aurora australis.

That might seem to be the end of the story – except that these days we have artificial satellites in orbit around the Earth and aeroplanes flying high above the level of the lower atmosphere. Both can be very adversely affected by sudden surges in the incidence of charged particles, subjecting delicate electronics and passengers alike to unhealthy doses of radiation. And this is not the only effect that mass ejections can have on the Earth, for they cause fluctuations in its own magnetic field. Such fluctuations, even if temporary, can have a devastating effect on man-made electrical grids. This was seen in 1989 when a massive X-ray flare was followed a few days later by a proton event of large magnitude. When these particles arrived at the Earth, they gave rise to a magnetic storm that shifted the Earth's field by as much as 8° in a few hours. This in turn produced large inductive currents in power lines, telephones and cable networks. In Canada, transformers were overloaded and large parts of the power network had to be shut down. It was thought that an event like this was rare and therefore of little significance. However, since then, careful study of the Sun has revealed that changes are afoot. Scientists and electrical engineers alike have learned a new respect for our Sun and what it is capable of.

A major leap forward in the study of solar behaviour was made possible by the launch, in December 1995, of the Solar and Heliospheric Observatory satellite – SOHO for short. This craft, which occupies a special orbit between the Sun and Earth, takes continuous pictures of the Sun's surface. Consequently, it has become a frontline tool for the observation of sunspots and therefore of the behaviour of the Sun's magnetic field. It has long been known that the Sun's polar magnetic field reverses its polarity every 11.1 years or so. It is this cycle that causes the rise and fall of sunspot activity in a similar cycle. However, careful study of

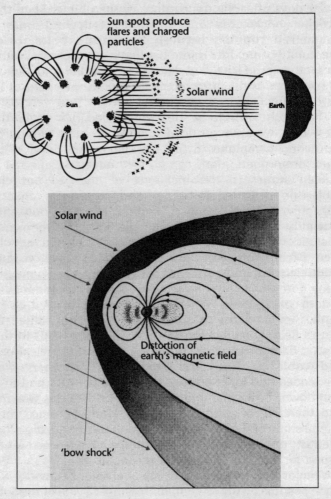

Figure 52: Solar wind and the Earth's magnetosphere

sunspot numbers reveals that this cycle is not the only factor involved.

The Sun reached one of its periods of maximum spot intensity in the middle of 2000. However, instead of a tailing off in activity following this, it formed another peak that was almost as strong in 2002. Since then, the number of spots has declined but not, it seems, their potential for causing mass ejections of solar plasma. In September of 2005, a spot called '798' fired off the largest X-class flare on record. What amazed scientists was not just the intensity of the burst but the fact that even though it appeared on the eastern edge of the Sun and was therefore not directed towards the Earth, it nevertheless produced a blackout of high-frequency transmissions and aurorae that were visible as far south as Arizona. This was not all. The spot which produced this flare carried on growing and, with never-before-witnessed activity, over the course of the next twelve days pumped out a further eight X-class flares.

Unfortunately, magnetic storms on the Sun cause more than aurorae and electrical blackouts. It is now recognised that solar activity has a profound effect on climate. We should therefore expect that when the sunspot cycle is close to a minimum, as it is now, our weather should be calm. However, in 2006 this does not seem to be the case. Even before the emergence of the X-class flares in September 2005, it was becoming clear that variations in our weather are becoming more extreme. This is generally attributed to 'global warming', a catch-all title that has unfortunately been hijacked by lobbyists as a useful weapon with which to attack governments, industry and consumers for making use of resources such as coal and oil, and thereby increasing the levels of 'greenhouse' gases such as methane and carbon dioxide. What is often forgotten, though, is that the 'greenhouse' effect of these gases is still an unproven theory.

While few will today dispute that the Earth is getting warmer – with possibly devastating effects upon nature's balance – it is not at all clear that the effluents of civilisation are the primary cause. If we look at the fossil record, it is clear that the Earth has at times been much colder than it is today and at other times much warmer. In fact, if we accept that the oil and coal which we extract from the ground is the residue of primeval forests, then there must have been a time before these forests appeared. At that time, virtually all the carbon present on Earth must have been in the form of 'greenhouse gases': either methane or carbon dioxide. This does not seem to have hindered the development of life or prevented cycles of global warming and cooling. It would seem, therefore, that human use of fossil fuels is only one of the factors behind climate change. There are other, possibly more important, ones that are beyond our control.

If we relate this back to what Arguelles and the Mayans have to say concerning solar cycles, then it would appear that the strange behaviour of the sunspot cycle is connected not just with global warming but with the changing of the ages. This is clearly linked with other 'signs in the sky' that, if read properly, are portents indicating the end of one age and the beginning of another. How these portents relate to prophecies contained in the Bible was the subject of my book *Signs in the Sky*. However, there are others that relate to Mayan cosmology and these too need to be considered.

AT THE END OF THE AGE

In *Signs in the Sky*, I drew attention to the way so many prophecies in the Bible pointed towards the year 2000 as the termination of one age and the start of another. My reasons for saying this concerned the presence of the Sun at the Orion star-gate on the day of the summer solstice, when all the planets were in close attendance. This I read as the symbolic meaning of the description given at the start of the Book of Revelation where 'The First and the Last, the Alpha and the Omega' stands amidst seven candlesticks, symbolic of the seven planets of the ancients.[6] The fourth chapter, following on from the symbolic description of Orion as alpha and omega, begins with the statement 'Lo, a door in heaven is opened'. I connected this with the symbolic 'opening of the star-gate' above the hand of Orion.

To witness this event in places most strongly connected with it, I led a tour group to first Egypt and then Israel. On 21 June, the day of the solstice itself, we witnessed the first part of the opening. I conducted a short ceremony at the foot of the Khafre Pyramid, the central one of the Giza group, which is symbolically linked with Alnilam, the central star of Orion's Belt. At the moment of the ceremony, the Sun was positioned exactly at the crossing point of the ecliptic and Milky Way, at the star-gate over the hand of Orion. Because of the angle of slope of the pyramid, it was also seen to apparently sit on top of it when directly in the east. At Jerusalem on 29 June, when the Moon joined the planets around Orion, we witnessed its dawn-rising over the Mount of Olives.

This was the second part of the 'opening' and, I believe, a very important moment in human history. For within weeks of this event, the Palestinian uprising began. It started right there, on top of the Temple Mount in Jerusalem, close to the Golden Gate where we watched the rising of first the Moon and then the Sun over the Mount of Olives. This uprising has continued to grow ever since, engulfing first Israel, then provoking the attack on the World Trade Center in New York, the invasion of Afghanistan and finally the Iraq War. Tens of thousands of people have already died in this 'War of Civilisations' and the conflict shows no sign of abating. If anything, it is getting worse and going global, with further attacks against Westerners in Bali, Madrid and London.

The perceived cause of this conflict is Islamic radicalism in the face of Western involvement in the Middle East. However, though this is undoubtedly the way it manifests itself, I do not believe this is the real root cause of what is going on in our world. All wars cause innocents to suffer but this war has an element that is in itself shocking in its callousness: the suicide bomber. For while people have in the past dropped bombs on civilian populations and even, in the case of Nazi Germany, made a serious attempt at the extermination of a whole race, they don't usually do this through religious motives. How can one account for young men, often well educated and of good family, walking into crowded restaurants and blowing themselves up? How does one account for suicide bombers killing people of their own religion (though of a different sect) even while they are at prayer in a mosque or at the graveside of a loved one? How can anyone imagine, least of all a devout Muslim, that God is going to approve of such an act that is unjust to those killed and brings so much suffering to the families left behind?

Such actions are beyond belief and seem quite inexplicable until one begins to realise that there is a cosmic dimension to what is happening in our world. I now believe that forces of chaos were unleashed on the world when, astronomically speaking, the Orion star-gate opened in the year 2000. This only began the process that is described in graphic and sometimes horrifying imagery in the Book of Revelation. We have much more to experience yet, for 'Lo, a door in heaven is opened' is only how Chapter 4 begins. It is not until Chapter 20 that an angel comes down from heaven with the key to the bottomless pit into which he throws the dragon, 'that ancient serpent, who is the Devil and Satan'.

The period we are in seems to be what in the Book of Revelation is described as the 'Tribulation', a word derived from the threshing of grain with a three-limbed whip (tribula) in order to separate the wheat from the chaff. Remembering this, I was more than interested to read a warning from the Mayan elders issued to humanity through Carlos Barrios via Mitch Battros. After some words concerning the recent tragedy in Guatemala itself, which had been badly hit by storms, he issued the following prophecy:

> Now is the time to take action, because we're warning, and we think everything is going to happen, and it is going to begin with water. Tsunami, or the water is going to touch all the continents, and create a lot of destruction to warning to the people, and second is coming earthquakes. The Mother Earth is beginning to move, and after this is the most dangerous. After this is fires who are coming, the magma. The big eruptions, like the possibility of the park, Yellowstone, and in other places around the world. This is just a demonstration of Mother Earth, but if we don't pay attention, if we don't stop the destruction, if we don't stop the pollution. OK. It's our responsibility, and millions of people can die . . .[7]

The apocalyptic prophecies of the Mayan elders, like those of the Book of Revelation, would be easier to dismiss as fantasies if we didn't have the evidence under our own noses that something is definitely afoot with our world. Revelation uses dream-like symbology that is hard to understand, yet if we look around us we can see that much of what is prophesied there is already happening. It predicts earthquakes and pestilence, wars and famine. We do not have far to look to see that this is indeed happening. Half of

Africa is starving and many countries there have been stripped of their adult populations by the scourge of Aids. The sudden tsunami in the Indian Ocean on 26 December 2004 produced huge devastation in countries as far apart as Sri Lanka and Thailand. Less than a year later, in September 2005, we saw the devastation wreaked on Louisiana by Hurricane Katrina. This was followed in October by an earthquake in Kashmir that left hundreds of thousands of people without food and shelter for the winter. How then can we say that these are empty prophecies? What more evidence do we need that something serious is happening with our world and it is not under our control?

I now believe that the Biblical prophecy of the opening of the bottomless pit, which is followed soon after (in Revelation 21) by the appearance of a 'New Heaven and a New Earth', is connected with the Mayan prophecy for the start of the fifth sun. As already discussed, on 21 December 2012, when the Mayan long count fulfils its 13th baktun since creation day on 13 August 3114 BC, the Sun will be positioned at the southern star-gate, where the ecliptic crosses the median plane of the Milky Way in the constellation of Sagittarius and over the sting of Scorpio. At midnight, the Sun will be aligned with the very centre of our galaxy. This will not be visible, however, as it will be night-time and the Sun will be on the opposite side of the Earth. However, dominating the southern sky will be the constellation of Orion, which, according to the Maya, was the place from which their father gods Hunahpu and Xbalanque came and which my esoteric reading of the Bible in *Signs in the Sky* reveals to be the sign of the Son of Man in heaven.

This moment, when the Sun is located at the southern star-gate and Orion, with its northern star-gate, is dominant in the night sky, will, I believe, signify the termination of the tribulation prophesied in the Book of Revelation and the true beginning of a new age. As a further sign – in the Mexico City area, at least – the shepherding star Arcturus, which both Edgar Cayce and J.J. Hurtak identify as the place from which teachers are sent to Earth, will transit close to the zenith about eight hours later. What will this mean for us in practical terms? It is hard to say, but perhaps it's worth noting that Mexico City is already a major centre for UFO activity. Huge numbers of sightings of either silvery objects

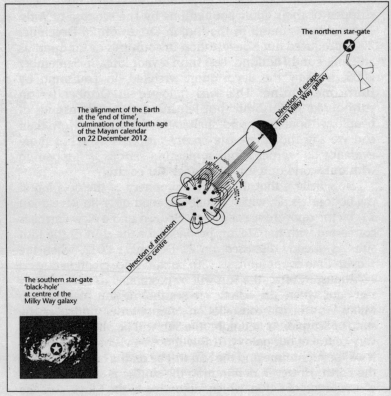

The northern star-gate

Direction of escape from Milky Way galaxy

The alignment of the Earth at the 'end of time', culmination of the fourth age of the Mayan calendar on 22 December 2012

Direction of attraction to centre

The southern star-gate 'black-hole' at centre of the Milky Way galaxy

Figure 53: Opening the second star-gate, on 22 December 2012

or lights in the sky have been reported by people and many times these have been recorded on video. There are eyewitness accounts from the area around Guadalupe, and, most telling of all, from an air-traffic controller at Mexico City International Airport.

Now, we should be careful not to read too much into this. However, given the strangeness of our times and the way that so many ancient prophecies are being fulfilled, both in terms of astronomy and in what's happening on the Earth, I think it behoves us to keep an open mind. The time is always later than we think. Let us be prepared, then, for major changes and accept that it is indeed the end of time as we have known it.

Epilogue

Over the course of the past ten years since *The Mayan Prophecies* was published, I have had many hundreds of people come up to me and ask me the same question: to where should they move in order to be safe. Unfortunately, I have no simple answer to this question. The fact remains that in the event of global changes, neither I nor anyone else can be sure exactly where is going to be safe. Like anyone else, I can only speculate on what seems a likely scenario.

That said, there are precautions that we can take and places that look, on the face of it, to be safer than others. It is over 60 years since Edgar Cayce said that, come the Earth changes, the climate of northern Europe would be changed in the blink of an eye. At the time he said this, it must have seemed fanciful in the extreme. However, we have more knowledge now than in his day of how important is the Gulf Stream to the preservation of an equable climate in northern Europe. As long as it continues to flow, it will remain possible to grow palm trees in northern Scotland and raspberries north of the Arctic Circle in Norway. Without it, the climate of England would be like that of Labrador in Canada: in winter, cold in the extreme.

We have known of the importance of the Gulf Stream for

many years now. However, it has recently been announced that, because of the accelerating melting of the Greenland ice sheet, the flow of warm water from the Gulf of Mexico is under threat. Melted ice is swelling the rivers of the Arctic Circle and diluting the northern seas with increasing amounts of fresh water. For the Gulf Stream to operate properly, dense, cold, salty water from the North Pole has to sink to the bottom of the Atlantic and flow southwards towards Mexico. Without enough salt in its composition, the cold northern water will not sink and play its part in forming the circulating current that allows hotter water from the Gulf to flow northwards.

The results of this impeded flow appear to be with us already. With warm water trapped in the Gulf of Mexico and the mid-Atlantic, temperatures have risen in both of these regions. Hot water equals energy, and this has to go somewhere. In 2005, it caused tropical storms of record intensity that affected the Caribbean islands, Mexico, Guatemala and the United States. Undoubtedly, we are going to see more of these to come, making these areas, as well as a cooling Europe, dangerous places in which to live.

This one example shows how difficult it is to predict just where the safe places are likely to be. The situation is made more difficult still if we take into account the possibility of unpredictable catastrophes such as major earthquakes, the eruption of super-volcanoes or even a collision between Earth and a comet or asteroid. All of these, were they to happen, could cause a doomsday scenario. The severity of such an event would determine whether or not the human race as a species would survive. It is entirely possible that we might not if the catastrophe were large enough.

In this worst-case scenario, speculation about how we should prepare and where might be the best place to be is entirely pointless. When the asteroid struck close to the Yucatán Peninsula (about 65 million years ago), it wiped out all the dinosaurs – not just those close to the point of impact but throughout the world. The reason for this is that the impact was so great it caused the Earth to be veiled with a thick cloud of dust and ash. This blocked out the sunlight and caused temperatures to plummet. The dinosaurs were unable to adapt to these new, harsher conditions and died either from the cold or, as likely, from starvation. The only members of their species to survive were the birds. These had already developed feathers and were, presumably, therefore able to keep warm enough to cope with the changes.

Now it is fairly pointless us trying, at an individual level, to plan for a contingency of this magnitude. However, there have been other, smaller events in world history that, while causing local hardship, have not caused the extinction of the human race. Around 75,000 years ago, a super-volcano exploded in the region of present-day Sumatra. This volcano left behind it a huge crater that has since filled with water to become Lake Toba. More importantly, it filled the upper atmosphere with sulphur dioxide that, when mixed in water, formed droplets of sulphuric acid. The effect of this was to cause a mini ice age that lasted for some 1,000 years. It is estimated that, in the course of this extended winter, around 60 per cent of the human population perished. However, just as importantly, 40 per cent survived. What we need to know is how to increase our chances of being among the 40 per cent of survivors, rather than the 60 per cent of victims.

A catastrophe of rather less magnitude occurred in the middle of the sixth century AD. The causes of this are still a matter of debate. Contemporary records indicate that fragments of a comet struck Britain, giving rise to the legend of the Wasteland. I wrote about this in *The Holy Kingdom,* which concerns a re-evaluation of the story of King Arthur and the Holy Grail. This book has proved to be highly controversial in Britain, and some academics disagree with its findings. However, there is no denying that there was a period of seven years in the mid-sixth century during which it was perpetual winter and the trees stopped growing. The effect of this catastrophe was to usher in the European Dark Age.

As I have already said, as there is no knowing where or when disaster is going to strike, it is impossible to cover every eventuality. However, it should be obvious that living in areas of earthquake or volcanic activity is courting disaster. It is also inadvisable to live somewhere like New Orleans or parts of London where the land is below sea level. Even assuming that the Mississippi levees are rebuilt and the

Thames Barrier made higher, we can be sure at some time the sea will overcome such obstacles. One day, these cities will be flooded just as sure as Atlantis lies at the bottom of the sea.

While we probably could not survive a major catastrophe such as the asteroid impact that wiped out the dinosaurs, we should prepare ourselves to meet lesser contingencies. We are more likely to be confronted by a cosmic winter lasting seven years than a wholesale ice age. As I see it, the key to survival is adequate, short-term food stocks. Fortunately, this is one thing that we can influence even if we have no control over the weather. Given the will, wherever we live – in Northern Europe, the Caribbean or Alaska – we can at least put by enough provisions to last us for a year or two. Every person reading this should, in my opinion, see about buying a sack or two of grain, some tinned food and dried pasta. Store this in the attic or spare room. Then, should doomsday come on 22 December 2012, you will be in a better position to survive at least the first phase of any catastrophe. What happens after that, when rations are running low, will be determined collectively. It will be down to communities rather than individuals to find means of self-support.

In the Dark Ages, the nuclei of such communities were the monasteries, where levels of organisation were such that agriculture, letters and religious instruction survived. We don't know what sorts of institutions will turn out to be best to meet the challenges of the future. What we do know is that in the immediate aftermath of a global catastrophe, such as the eruption of a super-volcano, chaos will ensue. Law and order will likely break down, and we will have the need of something akin to King Arthur's Knights of the Round Table to create areas of safety where civilisation can be preserved.

There is, however, more than physical survival to be considered. The Maya, like every other civilisation prior to the present, was founded on a metaphysical premise. They believed in a creator God (higher than the subordinate jaguar gods), in heaven and in hell. Expectation of life after death – whether as spirits or as physical people living on a different planet – partly explains the Mayans' willingness to make blood sacrifices: the souls of those who died would live on. The other reason, as we have seen, is that they wanted to ensure the sun kept rising, thus prolonging not only their own earthly lives but the life and prosperity of the land on which they lived. The Mayans therefore placed a great importance not only on their immediate, earthly existence but on what would live on after their deaths.

This attitude has lessons for us as we find ourselves living through what the Christian Bible calls the 'End Days' (as explained in detail in my earlier work *Signs in the Sky*). Rather than playing at Noah and trying to ensure our own survival, we need to ask ourselves some difficult questions concerning our own values and beliefs. Take a piece of paper and, before reading on, write your answer to this question: for what or for whom would you, in exceptional circumstances, be willing to lay down your life?

If you have children, then it is very likely you have written that you would be willing to sacrifice yourself for them. What, though, of your neighbour's children? Would you die for them too? And what of your country? Would you be willing to make the ultimate sacrifice in its defence? This is not to suggest that you should. It is simply an exercise to help you make yourself conscious of just where you stand on this issue.

Now let's take another example. Just suppose that an asteroid is even this minute heading towards planet Earth. Imagine that its orbit has been worked out by scientists using the most advanced computers and that they are certain it will hit the Earth, with devastating consequences, in five years' time. What would be your reaction to this news? How would knowledge that in five years' time we will all most probably die affect the way you live your life now? Again, take time to think about the question and write down your answer before proceeding further.

A common response might be: 'Well, if the world is going to end, then I may as well spend what money I've got on sex, drugs and rock and roll.' But is that really true? In reality, should the situation arise, I doubt that most people would actually act this way. Faced with the prospect of a dinosaur-type extinction in only five years' time, you wouldn't be fretting over trivia. On the contrary, aware that time is running out, you would be desperate to find immortality. Though you may not now think of yourself as a religious person, in these dire circumstances you would probably start praying to whatever God or gods you may at one time have believed in. This was certainly the case during the First and Second World Wars, which saw an upsurge of religious adherence.

This example gives us, I think, a clue as to how we should treat the Mayans' prediction of the end of time. Their claim that the present age of the world is going to come to an end in 2012 is the equivalent to the possibility of an asteroid strike. Under these circumstances, we owe it to ourselves to look deeper within. Since every culture in the past has held that man is not just a body but is also a soul, then we should try to understand just what this might mean and adjust our behaviour accordingly. On the other hand, if it turned out that the Maya were wrong in their expectations, then we would still have lost nothing: self-knowledge and a mature philosophy are assets worth having, whether or not we are facing a doomsday scenario.

All this is to look on the dark side of prophesied events. There is also, so we are told, much to look forward to. If we can but get through the predicted period of chaos (whatever its cause), we can expect a new golden age to emerge. It is my fervent hope and belief that we will then have wakened within us faculties that presently lie dormant. We will discover who and what we really are and see our planet brought back at last into harmony. That, at least, is the promise of the Bible, and it is one made to us whether we live in Kamchatka or Australia. Just how it will all pan out we can only guess. One thing we do know, though, is that 2012 is now just a few years away. We do not have long to wait to discover if 'were-jaguars' or any other kind of cosmic entities are going to come to our rescue. It is up to us to make sure that we are prepared for all eventualities.

Notes

PREFACE
1. What is meant here by 'star-gate' will be explained later.
2. The ecliptic is the pathway the Sun follows through the zodiac. It moves along this path at a rate of roughly 1° a day, making a full circuit in a year.
3. The Milky Way is a spiral galaxy shaped like a Catherine Wheel. Our Sun is one of millions that spiral around the centre of the galaxy, where astronomers believe there is a singularity or 'black hole'. The Sun is positioned near the end of one of the galaxy's spiral arms. We are therefore very far from the centre of the galaxy, which from our vantage we don't see as a spiral but rather as a stream or river of stars across the sky.
4. First published in 2000 by Transworld UK. New edition published in 2005 by A.R.E. Press, Virginia Beach, USA.

PROLOGUE
1. Published by Element Books in 1995.
2. A katun is a period of roughly 20 years: 20 x 360 days, to be precise.

317

CHAPTER 1: MEXICO REVISITED

1. The highest volcano in Mexico is Pico de Orizabo, an even larger strato-volcano near to Puebla, east of Mexico City.
2. Huitchilobos (Huitzilopochtli) was the Aztec god of war. He has many of the same characteristics as the Mayan 'Hero Twin' Hunahpu. Tezcatlipoca, or Smoking Mirror, was the Aztec equivalent of the Mayan god *K'awil*, about whom much more will be said later. (The more common spelling of both of these gods' names has been used here.)
3. Bernal Diaz del Castillo (trans. A.P. Maudslay), *The Discovery and Conquest of Mexico*, Kingsport Press, Inc., Kingsport, Tennessee, 1956.
4. The Aztecs seem to have always referred to Cortés by the name Malinche.
5. Del Castillo, *The Discovery and Conquest of Mexico*.
6. Close to the modern-day resort town of Tela on the Caribbean.
7. Interested readers are advised to read Michael Coe's book *Breaking the Maya Code* for a detailed history of Mayan decipherment.
8. Texcoco, which gave its name to the lake surrounding the Aztec capital, was a neighbouring and older city than Tenochtitlan. Today it is a suburb of Mexico City. The people of Texcoco were among the tribes who entered into federation with the Aztecs.

CHAPTER 2: THE CITY OF THE DEAD

1. *The Sign and the Seal* presented the case, implausible in my view, that the Ark of the Covenant was taken to Ethiopia and resides there to this day.
2. Giorgio de Santillana and Hertha von Dechend, *Hamlet's Mill*, Gambit Inc., Boston, 1969, p. 243.
3. See Mark Lehner, *The Complete Pyramids*, Thames & Hudson, London, 1997, p. 28.
4. Codices (or codex, in the singular) is the name given to native books, some of which were drawn before the arrival of the Spanish.
5. For this reason, it is today finding a use in surgical implements that help preserve life.
6. The zenith point is the portion of the sky that is directly

overhead. It is not to be confused with the astronomical north pole, which will only be seen coincident with the zenith if you are standing exactly on top of the terrestrial north pole.

7. José Diaz Bolio, *Why the Rattlesnake in Mayan Civilization*, Area-Maya, Merida, 1988, pp. 52–3.

CHAPTER 3: THE ROLLS OF TIME

1. As we will see later, some modern Mayans believe it to be linked to the cycles of the Pleiades star-group. It may have been linked to the cycles of Venus and Mars.
2. Friar Bernadino Sahagun, *A History of Ancient Mexico*, (trans. Fanny R. Bandelier, from the Spanish version of Carlos Maria de Bustamente), Fisk University Press, Nashville, 1932, pp. 108–9.
3. *Ibid.*, pp. 111–12.

CHAPTER 4: THE REDISCOVERY OF THE MAYA

1. J. Eric Thompson, *The Rise and Fall of Maya Civilization*, University of Oklahoma Press, Norman, Oklahoma, 1954 (revised 1966), p. 43.
2. *Ibid.*
3. It would appear that, while the Aztecs believed there are five ages and that we are living in the fifth age, for the Maya there are only four.

CHAPTER 5: DECODING THE MAYA LEGACY

1. J. Eric Thompson, *The Rise and Fall of Maya Civilization*, pp. 189–91.
2. Although I personally prefer the choice of 12 August 3114 BC, for the sake of consistency we are using a start date of 13 August 3114 BC. The advantage is that this is what Linda Schele adopted in her translations of the hieroglyphs on the Temple of the Cross group.
3. Friar Diego de Landa (trans. William Gates), *Yucatán Before and After the Conquest*, Dover Publications, New York, 1978, pp. 82–3.

CHAPTER 6: A NEW VISION OF THE MAYA CREATION

1. After further study since the publication of our book *The Mayan Prophecies*, I now regret the inclusion of these theories, which I think seriously undermined the credibility of other, more important, ideas concerning sunspot cycles and the Mayan calendar.

2. David Freidel, Linda Schele and Joy Parker, *Maya Cosmos*, William Morrow Quill, New York, 1993, p. 53.

3. Wägner and Macdowall, *Asgard and the Gods*, Swan Sonnenschein & Co., London, 1891, pp. 25–6.

4. Giorgio de Santillana and Hertha von Dechend, *Hamlet's Mill*, Gambit Inc., Boston, 1969.

5. Thompson, *The Rise and Fall of Maya Civilization*, p. 261.

6. Michael D. Coe, *The Maya Scribe and his World,* The Grolier Club, New York, 1973.

7. Justin Kerr, *The Vase Book: A Corpus of Rollout Photographs of Maya Vases*, Kerr & Associates, New York, 1980–89.

8. The *Paris Codex* is one of only four ancient Mayan books to have survived the autos-da-fé of the Spanish conquerors.

9. Such a scene is depicted on page 68 of the first volume of Justin Kerr's *The Maya Vase Book*. Very likely this is the image being referred to here.

10. De Landa, *Yucatán Before and After the Conquest*, pp. 60–1.

11. Thompson, *The Rise and Fall of Maya Civilization*, pp. 260–1.

12. See Chapter 6 in Linda Schele and David Freidel, *A Forest of Kings*, William Morrow Quill, New York, 1990.

CHAPTER 7: ASTRONOMY AND THE DEATH OF SEVEN MACAW

1. D. Goetz and S.G. Morley (English trans.), *The Popol Vuh, translated into Spanish by Adrian Recinos*, University of Oklahoma Press, 1950.

2. *Ibid.*

3. See Freidel, Schele and Parker, *Maya Cosmos*, p. 75.

4. The suggestion that the Maya knew of the Scorpio constellation was not a guess on Freidel's part. The same point is made in *Hamlet's Mill*, which was published in 1969 and is a classic on the cosmic meanings behind the world's myths. The authors refer to a Nicaraguan goddess called 'Mother Scorpion' and to the 'Old Goddess with the scorpion tail' of the Maya. They associate the goddesses

with Selket and Ishara tam.tim: scorpion goddesses of Egypt and Babylonia respectively. All of these goddesses they link with the constellation of Scorpio.

5. In the Mayan language, the plural form of a noun is given by adding '-ob', rather than '-s' as in English.

6. Susan Milbraith, *Star Gods of the Maya*, University of Texas Press, Austin, 1999, p. 274.

7. Just what the connection is between these caves and the Classic Mayan civilisation with its calendars is a matter for debate. However, when I visited the caves in 1998, I was shown Mayan hieroglyphs on the walls that are among the oldest yet discovered.

8. Anthony Aveni, *Stairways to the Stars*, Cassell, London, 1997, p. 74.

CHAPTER 8: JOURNEYS FROM THE EAST

1. Bernal Diaz del Castillo, *The Discovery and Conquest of Mexico* (trans. A.P. Maudslay), Kingsport Press, Inc., Kingsport, Tennessee, 1956, p. 72.

2. *Ibid.*, pp. 73–4.

3. Sourced from www.econ.ohio-state.edu/jhm/arch/coins/fallsoh.htm.

4. See A. Wilson and B. Blackett, *The King Arthur Conspiracy*, Trafford Publishing, Victoria, BC, Canada, 2005.

5. See Barry Fell, *America BC*, Pocket Books, Simon and Schuster, New York, 1989.

6. See *ibid.*, pp. 253–7.

7. *Ibid.*, p. 261. Extract copied by Fell from S.D. Peet's article in *The Mound Builders*, 1892.

8. See Thor Heyerdahl, *The Ra Expedition*, George Allen & Unwin Ltd., London, 1970.

CHAPTER 9: THE AZTECS AND ATLANTIS

1. Plato (trans. Thomas Taylor), Benjamin and John White, London, 1793, pp. 445–7.

2. *Ibid.*, pp. 450–1.

3. Hesiod (ed. and trans. Hugh G. Evelyn-White), *Works and Days*, 'The Homeric Hymns, and Homerica', Loeb Classics, Cambridge Mass., 1914, II: 109–201.

4. Ovid (trans. J. Dryden et al., ed. Sir S. Garth), *Metamorphoses in Fifteen Books, translated into English verse by the most eminent hands*, J. Tonson, London 1717.

5. Gregory Little and Lora Little, *The A.R.E.'s Search for Atlantis*, Eagle Wing Books Inc., Memphis, Tennessee, 2003, p. 156.

6. Details of their explorations are contained in their book, *The A.R.E.'s Search for Atlantis*.

CHAPTER 10: THE SERPENT CULT AND THE PLANETARY GODS OF THE MAYA

1. *The Plumed Serpent: Axis of Cultures*.

2. Several years later, I was sorry to hear that death had finally caught up with Don José. He died in October 1998, some seven months after I last saw him. I took his books back with me to England and tried, unsuccessfully, to find a publisher willing to pay to have them translated into English. Eventually, after that proved impossible, I did as he requested and in 2004 I donated them to the Library of Anthropology at the British Museum in London.

3. De Landa, *Yucatán Before and After the Conquest*.

4. In Classic times, the second zenith transit of the Sun, which took place on 16 July at Chichen Itza, was on 2 August at Palenque.

CHAPTER 11: AT THE END OF THE AGE

1. For much more information on the subject of the star-gates, please see my book *Signs in the Sky*, published by Transworld, London, 2000.

2. See Robert Bauval and Adrian Gilbert, *The Orion Mystery*, Heinemann, London, 1994, for details.

3. The so-called Cerne Abbas Giant is evidence of an Orion cult in Dorset that probably goes back to the Iron Age. A chalk figure cut into the turf of a hillside, it is some 100 yards tall and depicts the constellation of Orion. Its priapic nature is seen by some to indicate a close connection with fertility cults but in my opinion it indicates that Orion symbolises the father principle.

4. '7/15/42 Eula Allen's Life Reading 2454–3', contained on

the CD-ROM *The Totally New Complete Edgar Cayce Readings*, All Edgar Cayce LLC, 2002.

5. *Ibid.*, reading 5749–14.

6. The so-called 'Book of Enoch' is a non-canonical work of Gnostic Jews that goes back to at least the time of Jesus. Fragments of the work have been found among the Dead Sea Scrolls but the whole has been preserved, in translation, in Ethiopia. One such version now resides in the Bodleian Library in Oxford, England, and has been translated. It concerns the adventures of the Old Testament patriarch Enoch, his soul-journeys through the heavens, his meetings with such angels as Michael and Gabriel and how God instructed him to tell the Nephalim (or 'fallen angels' who, on coming to Earth, had perverted the course of human evolution) that they would be imprisoned. A controversial work with some similarities to the 'Book of Revelation', the 'Book of Enoch' is referred to in the New Testament and was evidently known to Jesus himself.

7. J.J. Hurtak, *The Keys of Enoch*®, Academy for Future Science, Los Gatos, California, USA, 1987, p. 595.

8. According to my detailed researches on the subject, the Crucifixion took place on 15 April AD 29, the Resurrection two days later, on 17 April, and the Ascension 39 days after that, on 27 May AD 29. On this day, the Sun was standing exactly at the Orion star-gate, at the crossing of the ecliptic by the Milky Way. How I arrived at these dates is explained in my book *Signs in the Sky*.

9. Hurtak, *The Keys of Enoch*®, p. 54.

10. *Ibid.*, p. 61.

11. *Ibid.*, p. 181.

12. *Ibid.*, pp. 181–2.

CHAPTER 12: CYCLES OF THE SUN

1. Though I may have spelt it incorrectly, I have transcribed the word Sikim from what Carlos says on the video I recorded. I presume it means 'eagle' and is drawn from Mam or some other Mayan language from Guatemala rather than Mexico. In fact, we were at Cahal Pech on 16 June 2005 and this, on the Mayan 260-day Tzolkin count, is called 2 *Men*, with Men meaning eagle in Yucatecan

Mayan. I therefore assume this is what he was referring to.

2. I later discovered that such a bag is part of the regalia of a Mayan shaman. It would have contained crystals and other objects of spiritual value. These were not revealed to us but kept concealed within the bag.

3. At the time of writing, this article was located on the Internet at

www.trans4mind.com/counterpoint/barrios.shtml.

4. José Arguelles, *The Mayan Factor*, Bear & Co., Santa Fe, 1987, p. 109.

5. It is called the 'Kolob' by J.J. Hurtak.

6. See my earlier book, *Signs in the Sky*, for a detailed astronomical interpretation of the first chapters of the Book of Revelation.

7. From the transcript of an interview of the Mayan elder Carlos Barrios by Mitch Battros. At the time of writing, this interview was on the Internet at

www.mayanmajix.com/art2094.html.

Bibliography

Alexander, Hartley Burr, *Latin-American Mythology*, New York, 1964

Arguelles, José, *The Mayan Factor*, Bear & Co., Santa Fe, 1987

Aveni, Anthony, *Skywatchers of Ancient Mexico*, University of Texas Press, Austin, 1980
Stairways to the Stars, Cassell Publishers Ltd, London, 1997

Bauval, Robert and Adrian Gilbert, *The Orion Mystery*, Heinemann, London, 1994

The Book of Chilam Balam of Chumayel, first published by the University of Pennsylvania, 1913, reprinted with foreword by G.B. Gordon, Aegean Park Press, Laguna Hills, California, 1993

Calleman, Carl Johan, *The Mayan Calendar: Solving the Greatest Mystery of our Time*, Garev Publishing International, US, 2001

Cayce, Edgar Evans, *Edgar Cayce on Atlantis*, Warner Books, New York, 1968
The Totally New Complete Edgar Cayce Readings (CD-ROM), All Edgar Cayce LLC, 2002

Chadwick, John, *The Decipherment of Linear-B*, Penguin Books, London, 1961

Childress, David H., *Lost Cities of North and Central America*, Adventures Unlimited Press, Stelle, Illinois, 1992

Coe, Michael D., *Breaking the Maya Code*, Penguin Books, London, 1994

Reading the Maya Glyphs, Thames & Hudson, London, 2001

The Maya, Thames & Hudson, London, 1993

Coe, Michael D. and Justin Kerr, *The Art of the Maya Scribe*, Harry N. Abrams Inc., New York, 1998

Darwin, Charles, *The Origin of Species by Means of Natural Selection*, Penguin, London, 1982

De Landa, Diego, *Relación de las Cosas de Yucután*, Mexico City, 1959

Yucután Before and After the Conquest, Dover Publications, New York, 1978

De Santillana, Giorgio and Hertha von Dechend, *Hamlet's Mill*, Gambit Inc., Boston, 1969

Del Castillo, Bernal Diaz (trans. A.P. Maudslay), *The Discovery and Conquest of Mexico*, Kingsport Press, Inc., Kingsport, Tennessee, 1956

Del Rio, Antonio, *Description of the Ruins of an Ancient City discovered near Palenque*, London, 1822

Diaz Bolio, José, *The Geometry of the Maya*, Area Maya, Merida, 1987

The Mayan Natural Pattern of Culture, Area Maya, Merida, 1992

Why the Rattlesnake in Mayan Civilization, Area Maya, Merida, 1988

Donnelly, Ignatius (eds Sykes, Egerton), *Atlantis the Antediluvian World*, Sidgwick & Jackson, London, 1970

Fell, Barry, *America BC*, Simon & Schuster Pocket Books, New York, 1989

'The Comalcalco Bricks: Part 1, the Roman Phase', *The Epigraphic Society Occasional Papers*, Volume 19

Freidel, David, Linda Schele and Joy Parker, *Maya Cosmos*, William Morrow Quill, New York, 1993

Gallenkamp, Charles, *Maya*, Viking/Penguin, New York, 1985

Gann, Thomas, *Mystery Cities of the Maya*, Gerald Duckworth & Co., London, 1925

Gates, William (trans.), *Friar Diego de Landa's Yucatán Before and After the Conquest*, originally published by The Maya Society of Baltimore, 1937, reprinted by Dover Books, New York, 1978

Gilbert, Adrian, *Signs in the Sky*, Transworld, London, 2000

Gilbert, Adrian and Maurice Cotterell, *The Mayan Prophecies*, Element Books, Shaftesbury, 1995

Gilbert, Adrian, Alan Wilson and Baram Blackett, *The Holy Kingdom*, Bantam, London, 1998

Goetz, D. and S.G. Morley (English trans.), *Popol Vuh, translated into Spanish by Adrian Recinos*, University of Oklahoma Press, 1950

Grant, Michael and John Hazel, *Who's Who in Classical Mythology*, J.M. Dent, London, 1993

Hancock, Graham, *Fingerprints of the Gods: A Quest for the Beginning and the End*, Heinemann, London, 1995
The Sign and the Seal: A Quest for the Lost Ark of the Covenant, Heinemann, London, 1992

Heyerdahl, Thor, *The Ra Expedition*, George Allen & Unwin Ltd, London, 1970

Hurtak, J.J., *The Keys of Enoch®*, The Academy for Future Science, Los Gatos, California, 1987

Jenkins, John Major, *Maya Cosmogenesis 2012*, Bear & Co., Santa Fe, 1998

Kerr, Justin, *The Vase Book: A Corpus of Rollout Photographs of Maya Vases*, Kerr & Associates, New York, 1980–9

Krupp, E.C., *Echoes of the Ancient Skies*, Oxford University Press, Oxford, 1983

La Violette, Dr Paul, *Earth Under Fire*, Starlane Publications, 1989

Lehner, Mark, *The Complete Pyramids*, Thames & Hudson, London, 1997

Little, Gregory and Lora Little, *The A.R.E.'s Search for Atlantis*, Eagle Wing Books, Memphis, 2003

Maudsley, A., *Archaeology, Biologia Centrali Americana*, London, 1889–1902

Milbraith, Susan, *Star Gods of the Maya*, University of Texas Press, Austin, 1999

Morley, Sylvanus Griswold, *An Introduction to the Study of the Maya Hieroglyphs*, originally published by the Government Printing Office, Washington D.C., in 1915, reprinted by Dover Publications, New York, 1975

Muck, Otto, *The Secrets of Atlantis*, Collins, London, 1978

Ovid (trans. J. Dryden et al., ed. Sir S. Garth), *Metamorphoses in Fifteen Books, translated into English verse by the most eminent hands*, J. Tonson, London 1717

Plato (trans. Thomas Taylor), Benjamin and John White, London, 1793

Recinos, Adrian and Delia Goetz (trans.), *The Annals of the Cakchiquels and Title of the Lords of Totonicapan*, University of Oklahoma Press, Norman, 1953

Robicsek, Francis, *The Maya Book of the Dead: The Ceramic Codex*, University of Virginia Art Museum, Charlottesville, Virginia, 1981

Sahagun, Friar Bernadino (trans. Fanny R. Bandelier), *A History of Ancient Mexico*, Fisk University Press, Nashville, 1932

Schele, Linda and David Freidel, *A Forest of Kings*, William Morrow Quill, New York, 1990

Schele, Linda and Peter Mathews, *The Code of Kings*, Simon & Schuster Scribner, New York, 1998

Schele, Linda and Mary Ellen Miller, *The Blood of Kings*, Kimbell Art Museum, Fort Worth, 1986

Stephens, John L. and Frederick Catherwood, *Incidents of Travel in Central America, Chiapas and Yucatán* (2 vols), first published by Harper & Brothers, New York, 1841, reprinted by Dover Publications, New York, 1969
Incidents of Travel in Yucatán (2 vols), first published by Harper & Brothers, New York, 1843, reprinted by Dover Publications, New York, 1963

Taube, Karl, *Aztec and Maya Myths*, British Museum Press, London, 1993

Tedlock, Barbara, *Time and the Highland Maya*, University of New Mexico Press, Albuquerque, 1982

Tedlock, Dennis (trans.), *Popol Vuh*, Touchstone (Simon & Schuster), New York, 1986

Thompson, J. Eric, *The Rise and Fall of Maya Civilization*, University of Oklahoma Press, Norman, 1966

Thurston, Hugh, *Early Astronomy*, Springer Verlag, New York, 1994

Tompkins, Peter, *Mysteries of the Mexican Pyramids*, Thames & Hudson, London, 1987

Van Auken, John and Lora Little, *The Lost Hall of Records*, Eagle Wing Books, Memphis, 2000

Von Däniken, Erich, *Chariots of the Gods? Unsolved Mysteries of the Past*, G.P. Putnam's Sons, London, 1987

Villacorta, Carlos and J. Antonio Villacorta (eds), *The Dresden Codex*, originally published in Guatemala, 1930, reprinted by Aegean Park Press, Walnut Creek, date unknown

Wägner, Dr. W. and M.W. Macdowall, *Asgard and the Gods*, Swan Sonnenschein & Co., London, 1891

Wauchope, R., *Modern Maya Houses: A Study of Their Archaeological Significance*, Carnegie Institution of Washington, 1938

Wilson, A. and B. Blackett, *The King Arthur Conspiracy*, Trafford Publishing, Victoria, BC, Canada, 2005

Index